Primary Energy

Present Status and Future Perspectives

Edited by
Klaus O. Thielheim

With 224 Figures, Some in Color

Springer-Verlag
Berlin Heidelberg New York 1982

Professor Dr. K.O. Thielheim
Institut für Reine und Angewandte Kernphysik
Universität Kiel, Olshausenstr. 40–60
2300 Kiel, FRG

ISBN 3-540-11307-X Springer-Verlag Berlin Heidelberg New York
ISBN 0-387-11307-X Springer-Verlag New York Heidelberg Berlin

Library of Congress Cataloging in Publication Data. Main entry under title: Primary energy.
Bibliography: p. Includes index. 1. Power resources--Addresses, essays, lectures.
2. Power (Mechanics)--Addresses, essays, lectures. I. Thielheim, Klaus O.   TJ163.24.P74
333.79   82-831 AACR2

Printing and bookbinding: Beltz, Offsetdruck, Hemsbach/Bergstr.
2132-3130-543210

Preface

The enormous public interest of specialists as well
as of engaged and concerned citizens in the energy
problem can be understood in view of the fact that the
future of national and world-wide economy depends on
the availability of sufficient primary energy. The
questions arising are: which forms of primary energy
exist principally? by what means and at what cost can
they be brought to useful application? and what is
their possible role in the present and future energy
scenario?

Another reason which may not be so obvious, but which
eventually may prove to be of great importance as far
as public acceptance of energy technologies is concerned, lies in the fact that the existing conscious
or subconscious fears arising from confrontation with
scientific and technological progress — to which even
for the educated layman intellectual access is difficult — have been sublimated onto the energy problem
and especially onto the problem of nuclear energy. Unlike other developments, the emergence of nuclear energy has brought to our notice the ambivalence of ad-
vancing science and technology, which may either be
used peacefully or misused militarily.

Nuclear energy can help to overcome the increasing
hunger for energy in the world, but it can also lead
to the extinction of human life from the surface of
this plant. More and more, mankind is confronted
with chances and risks of new discoveries. Scientific
and technological innovations are appearing at shorter
and shorter intervals. Since the beginning of the industrial revolution they have become infinitesimally
small in comparison with time scales of biological
evolution. But now they are even becoming short in
comparison with the span of one generation. This makes
it difficult to keep pace with the changes in the
ambiente to which we are all subject. The impact of
these innovations on the individual and on the communities is increasing, along with the risks involved.
Will the ability for self-organization prove to be
sufficient for the human race to adapt itself to the
conditions of accelerating progress and to maintain
the equilibrium of economic, political, and social
structures?

It is my feeling that the responsibility imposed on
the scientist who has insight into some of these developments also includes the obligation to transfer
information relevant to these problems to the broad
majority of people interested and concerned, to enable them to form their own opinion of the necessi-
ties, chances, and risks associated with the energy
problem.

For this reason, I organized a series of lectures by
outstanding specialists working on the various aspects of the energy problem: these have aronsed such
great interest that I was encouraged to ask these
contributors to make their manuscripts available to
a larger public. In order to complete the spectrum
of subjects covered, I additionally asked some colleagues to write articles on their own specialized
subjects. Thus I hope that the present volume may be
useful in delivering first-hand information not only
to specialists but also to the engaged layman. It is
my concern to acknowledge gratefully the work under-
taken by these colleagues in spite of their many
other obligations. I also wish to thank the Springer
editors, who with great experience and patience,
were prepared to publish this book.

Kiel, February 1982   K.O. Thielheim

Contents

The Physical Concept of Energy

K. O. Thielheim[1]

1 Use of Energy in the Prescientific Era

Man used nuclear energy from the sun long before the physical concept of energy became known. In the form of electromagnetic radiation solar energy travels through interplanetary space, in the
form of chemical energy it is stored in organic materials on the
surface of the earth. Collected over large areas and through periods of months in plants, solar energy is transformed into muscular power of man and animals. By integration through decades in
wood and through millions of years in coal solar energy provided
the fuel for steam engines, the last energy technology of the prescientific era. Solar energy is transformed into kinetic energy
of air and thus may drive windmills, it is also transformed into
potential energy of water and thereby may power water mills.

In the seventeenth century people began to think of a machine able
to move perpetually and still perform useful work. They thought
of water flowing from a higher to a lower level driving an Archi-
medes' screw and hence lifting simultaneously the same amount of
water to its original level. The impossibility of such a device
could not be obvious to the technicians of those times. They were
not conscious of the existence of the physical quantity now called
energy and of its properties. Very much like the search of alche-
mists for gold to be made from non-precious materials, the search
of energy technicians in the prescientific era for a perpetuum
mobile was in vain.

2 The Discovery of Mechanical Energy

The systematic investigation of nature started with Galileo Gali-
lei's experiments. He found the law of bodies falling or sliding
on an oblique plane. Then Christian Huygens found the law of the
swinging pendulum. In his notes we find a mathematical expression
for a physical quantity which we would call kinetic energy today.
But he did not recognize the fundamental importance of his discovery.

1 Institut für Reine und Angewandte Kernphysik, Universität Kiel,
  Olshausenstr.40-60, 2300 Kiel, FRG

Fig. 1. The swing (by Nicolas Lancret)

Instead of a pendulum we wish to consider a swing as is shown in
the beautiful painting of Nicolas Lancret reproduced in Fig.1.
We will neglect the effects of friction in the air and in the suspension and observe the height h above the ground and the velocity v. Both quantities — h as well as v — change their values perpetually while the swing is moving to and fro. At its turning
points the direction of motion changes, the velocity being equal
to zero for just one moment. At its lowest point the velocity has
its maximum value. But, although height and velocity are changing
continuously, there exists a combination of both quantities which
does not change with time but keeps a constant value. This combination is named the mechanical energy of the system.

The mechanical energy turns out to be the sum of two contributions
one of which, the potential energy, is essentially determined by
the height h above the ground, while the other contribution, the
kinetic energy, is essentially determined by the square of the
velocity. Also potential and kinetic energy respectively change
their values continuously while the swing moves. But the sum of
both contributions, the total mechanical energy, is a conserved
quantity.

One can make use of the potential energy of bodies above ground
to define a unit of energy. Confining ourselves to the basic procedure at the expense of precision, we may consider a weight of

1 kg lifted from the plate of the table by about 10 cm. The potential energy of the weight thereby is increased by 1 J.

It is an astonishing historical fact that neither Kepler, who found the laws governing the motion of planets, nor Newton, who found the law of gravitation, discovered the law of conservation of mechanical energy. Gottfried von Leibniz recognized the importance of the notion of kinetic energy, which in 1686 he named vis viva. In the year 1717 the word energy occurs for the first time in a letter by Johann Bernoulli who applied this name to a physical quantity now called virtual work. But the general formulation of the law of conservation of mechanical energy was not found before 1748 by his son Daniel Bernoulli, and independently in 1743 by Jean Laurant.

3 Heat

Let us have another look at the swing. When it has been pushed it will move for a certain time and then finally come to rest. Seemingly mechanical energy has got lost. Actually it has been transformed into heat energy under the influence of friction.

For a long time controversial ideas existed about the nature of heat. Some people meant that heat is a kind of substance. In 1789 the chemist Antoine Lavoisier named this substance caloricum. Other people, like Daniel Bernoulli in 1738, suspected that heat has something to do with the motion of the smallest particles of matter. This controversy still existed when in 1798 in the arsenal at Munich Count Rumford when drilling guns made the observation that practically an unlimited amount of heat may be produced by friction. This appeared to be a very serious argument against the hypothesis of heat being a substance.

A change in the understanding of the nature of heat occurred when in 1840 in Rotterdam Robert Julius Meyer, then 27 years old, went aboard a ship bound to Southeast Asia. He carried with himself Lavoisier's treatise on chemistry wherein the latter put forward the hypothesis that heat inside living bodies originates from the slow oxidation of food. Meyer connected this idea with his own observations that in tropical zones the venous blood of his patients exhibited a brighter color. He concluded that due to higher outside temperatures and the thereby decreased emission of heat from the body only a reduced production of heat inside the body and therefore a decreased consumption of oxygen had taken place. On this basis Meyer pursued the idea that the energy produced inside the human body by oxidation of food is balanced by the energy this body gives away as heat to the surrounding air or as mechanical work performed by the muscles. Thus he intuitively conceived a general law of conservation of energy, including heat and mechanical energy, which is called the first law of thermodynamics nowadays.

Robert Julius Meyer was also able to determine the mechanical heat equivalent, i.e., the conversion factor of mechanical to heat energy. In 1845 this factor was measured more accurately by James Prescott Joule after whom the unit of energy is named today. The mechanical energy of 4186 J corresponds to a heat energy of 1 kcal, which is the amount of energy necessary to heat 1 kg of water by one degree. In therms of these conventional units, a large mechanical effort is necessary for the production of a small amount of heat. In order to verify this you may try to warm up the water in your bathtub by stirring when it becomes cooler. If there is any noticeable effect at all, it will be due to the production of heat inside your body as a consequence of its activity.

The first law of thermodynamics was not yet known when in 1824 Sadi Carnot considered the question whether heat energy may be transformed completely into mechanical energy. He based his considerations on a principle, today called the second law of thermodynamics, stating that heat is transferred always from bodies of higher temperature to those of lower temperature. In his gedankenexperiment Carnot considered an ideal circular process between two levels of temperature, explained in Fig.2. At a lower temperature T_1 a gas is compressed while heat energy is transferred to the surrounding medium. Then, at a higher temperature T_2 the same gas expands while receiving heat energy. Between these two levels the gas is brought from the lower temperature to the higher temperature by compression, while the exchange of heat energy is inhibited by an insulation. Correspondingly, the gas is brought from the higher temperature to the lower temperature level by expansion, while again the exchange of heat energy is inhibited. From the analysis of this idealized heat engine Carnot concluded

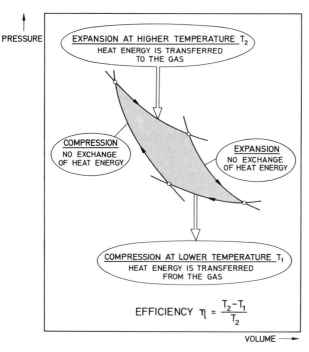

TRANSFORMATION OF HEAT ENERGY
INTO MECHANICAL ENERGY
CARNOT'S PROCESS

PRESSURE

EXPANSION AT HIGHER TEMPERATURE T_2
HEAT ENERGY IS TRANSFERRED
TO THE GAS

COMPRESSION
NO EXCHANGE
OF HEAT ENERGY

EXPANSION
NO EXCHANGE
OF HEAT ENERGY

COMPRESSION AT LOWER TEMPERATURE T_1
HEAT ENERGY IS TRANSFERRED
FROM THE GAS

EFFICIENCY $\eta = \dfrac{T_2 - T_1}{T_2}$

VOLUME ⟶   Fig.2

that a complete conversion of heat energy to mechanical energy is impossible. Part of the heat energy is always lost to the surrounding medium. The efficiency — that is the ratio of output mechanical work to input heat energy — in real heat engines is always smaller than in the ideal device. For example, James Watt's heat engine had an efficiency of only approximately 3%, while modern high-temperature steam engines have an efficiency of more than 30%.

4 Transport of Energy in the Form of Electricity

Heat engines first converted chemical energy from wood, later from other fossil fuels like coal, oil, and gas, to heat and further to mechanical energy. They substituted as an army of technical slaves the muscular power of man and animals. At that stage mechanical energy was available only where the heat engine was working.

About the middle of the last century the mysteries of electricity were disclosed by the discoveries of Michael Faraday, James Clark Maxwell, and many others. They led to the invention of electric generators and motors so that energy could be transported to any place where it was needed. Today the continents are cross-crossed by a network of electric transmission lines. In the Federal Republic of Germany for example, in 1972 this network carried as much as 1 billion billion Joule.

Looking at these transmission lines it is interesting to note that actually the energy is not transported inside the wires but, guided by the electric current, it is flowing through the surrounding space.

5 Radiation

But energy can also be transported through empty space disconnected from matter in the form of light or heat radiation, as from the sun. Still, there are other types of electromagnetic radiation distinguished from light and radiative heat by different wavelengths. The electromagnetic spectrum comprises gamma radiation emitted by atomic nuclei at a wavelength around 10 times a millionth of a million centimeter, optical light emitted by atomic electrons at a wavelength around 100 times a millionth centimeter, and radio frequency signals radiated from antenna in the range between centimeters and hundreds of meters.

There are also other forms of radiation different by their properties from electromagnetic radiation. One, which has been observed just recently, reveals its existence in the form of extremely weakly interacting massless particles called neutrinos. Another one, which has not yet been observed but is predicted by Einsteins theory of general relativity and now is being searched for intensively, is gravitational radiation.

6 Fundamental Interactions and Binding Energy

Let us once again have a look at the swing. As was said, its potential energy is essentially determined by its distance from the ground. The larger the distance, the greater the potential energy. Actually we are dealing with a system consisting of two parts, one is the body on the swing, the other is the earth, both being connected by the force of gravity. To this gravitational interaction corresponds a binding energy. Each system composed of different objects can principally give away energy when it is transferred from a state of weaker binding into a state of stronger binding. If for example the ropes of the swing break and the person sitting on it falls down to the gound, the available binding energy is first transformed into kinetic energy and then — painfully — further into deformation energy.

Figure 3 gives some more examples of systems composed of two objects. One is the atomic nucleus of deuterium consisting of a proton and a neutron. These constituents are bound together by the nuclear force. This is the strongest of the four fundamental interactions known to us but at the same time it has a very short range. The strong interaction governs the structure of the microcosmos, of the nuclear and subnuclear objects like nuclei and atomic particles.

The hydrogen atom is another example. It consists of a proton and an electron, bound together by electromagnetic interaction. This well-known force governs the structure of objects like atoms and molecules, liquids and solid bodies, the world of chemistry.

The binary planet of earth and moon is still another example. Very much like the system consisting of the earth and the person on the

THE FUNDAMENTAL FORCES OF NATURE

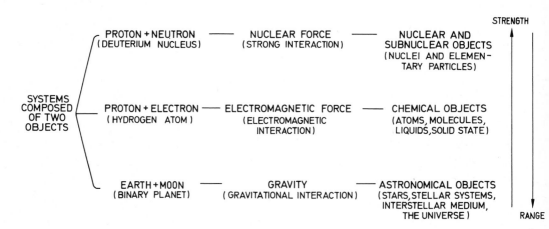

A FOURTH FUNDAMENTAL FORCE IS CONSTITUTED BY THE WEAK INTERACTION, WHICH IS RESPONSIBLE FOR THE BETA DECAY OF ATOMIC NUCLEI.

Fig.3

swing they are bound together by gravitational force. This gravitational interaction governs the structure of astronomical objects like stars, galaxies, interstellar gas and dust clouds, and finally the structure of the universe as a whole. Gravitation is a very weak interaction with a very long range.

To complete the list of fundamental interactions we have to mention the so-called weak interaction. It can be made responsible for certain forms of radioactive decay, called beta decay and is related to the existence of neutrinos.

7   Nuclear Energy

Nuclear energy becomes available when the constituents of nuclei, the nucleons, are transferred from a state of weaker binding into a state of stronger binding. The diagram in Fig.4 explains how this is possible. The horizontal scale refers to the number of nucleons within a nucleus, while the vertical scale indicates the binding energy per nucleon inside the nucleus.

An appropriate unit of energy in the nuclear region is 1 MeV, which is approximately equal to 1/6 of a millionth of a millionth Joule. Obviously the amount of energy released in the individual reaction is very small, around 201 MeV per fission and close to 18 MeV per fusion in the case of the deuterium-tritium reaction, discussed later.

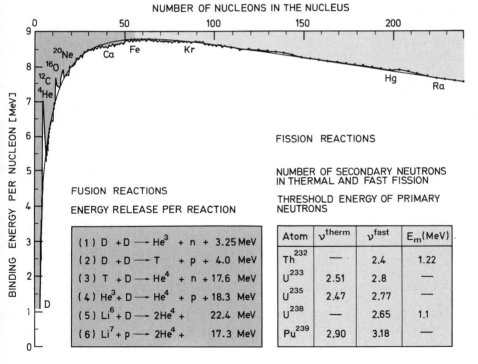

Fig.4

8

The nuclear binding energy liberated in the complete fission of
1 g uranium 235 amounts to 81 billion Joule. This may be compared
with the chemical binding energy of about 30,000 J set free in
the oxidation of 1 g of coal.

As may be seen from the diagram, the binding energy of the nucleons is strongest in an intermediate range of atomic weight around
iron. The binding is comparatively weak in the range of lighter
atoms and also in the range of heavier atoms. Thus there are two
ways leading from a state of lower binding to a state of stronger
binding of nucleons: one through fusion of lighter nuclei, the
other through fission of heavier nuclei. On both ways nuclear
binding energy is set free.

The empirical dependence of the binding energy per nucleon on the
number of nucleons in a nucleus may be interpreted by considering
the fact that the nuclear force has an extremely short range and
therefore is effective practically only between nuclei which are
adjacent to each other. Neglecting surface effects, one would thus
expect the binding energy per nucleon to be independent of the
size of the nucleus. In that case the diagram would show just a
horizontal line, with all nucleons already being in the state of
strongest binding. But taking into account the surface effects,
it is obvious that those nucleons which are positioned nearer to
the surface have around themselves fewer neighbors than other nucleons further inside the nucleus such that the former are weaker
bound than the latter. Since the ratio of surface to volume increases with decreasing volume, the binding energy per nucleon becomes smaller on the way from intermediate to light nuclei.

Furthermore, besides the short-range attracting nuclear forces
there are repelling long-range electrostatic forces acting between
the positively charged nucleons, which is also effective between
the more distant protons inside the nucleus. As the number of protons increases together with the number of nucleons, the binding
energy per nucleon also becomes smaller on the way from intermediate to heavier nuclei.

8 Nuclear Fusion

Due to the short range of nuclear forces which give rise to the
fusion reaction, the nuclei in question have to have a kinetic energy well about a "critical" energy, in order that they can overcome the long-range repulsion by electrostatic forces.

But even then the probability of scattering of the two nuclei is
much stronger than the probability of fusion, hence the particles
scatter each other several times before reacting. Therefore, a
large-scale energy release through fusion is possible only in a
medium whose kinetic heat energy is so high that the corresponding temperature exceeds a certain critical ignition temperature.

Additionally, according to what is called Lawsons's criterion,
the plasma must be kept at a sufficiently high density for a sufficiently long period of time so that the energy gain through fusion exceeds the energy loss through electromagnetic radiation.

In the sun and other fixed stars the confinement of the plasma is provided for by the gravitational field evoked by the enormous mass of these stellar objects. In the center of the sun at a temperature of about 15 million degrees the so-called proton-proton process prevails among other fusion reactions. Deuterons are formed in proton-proton encounters. Then helium-3 is produced in proton-deuteron collisions. Finally, encounters between two helium-3 nuclei result in the production of one helium-4 nucleus and two protons. The net effect is the production of helium-4 through hydrogen fusion.

At about 10 million degrees the so-called carbon cycle starts to contribute, in which with the help of carbon-12 as a catalyst protons are melted together to form a helium-4 nucleus, as is explained in some detail in Fig.5. At still higher temperatures helium-4 nuclei can combine to produce even heavier elements.

The time scales involved in stellar fusion reactions exclude their use in fusion reactors. Some reactions which are of interest in view of a possible technical application are listed at the bottom left of Fig.4. The most promising candidate is the tritium-deuterium reaction since its ignition temperature is "only" about 50 million degrees.

Deuterons are the nuclei of heavy hydrogen, of which one is present among 125 protons. The oceans are estimated to contain about 100 million million tonnes of deuterium, so that it is available in practically unlimited amounts all over the world.

Tritium decays with a half-life of only about 12.3 years and therefore does not occur in nature. It can be produced by neutron capture in natural lithium, which is quite abundant in the crust of

FUSION IN STARS
CARBON CYCLUS

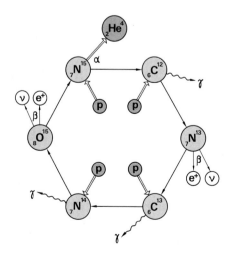

Fig.5

the earth and consists of about 7.4% of lithium-6 and 92.6% of lithium-7. Either in fission of fusion reactors lithium-6 by absorption of a neutron can be transformed into helium-4 and tritium in an exothermal reaction, while lithium-7 by neutron capture is converted into helium-4, tritium, and an extra neutron in an endothermal reaction.

About 20% of the energy released during the thermal fusion process is kinetic energy of helium-4 nuclei, α-particles, which are intended to deposit their energy within the plasma. About 80% of the energy is taken away by secondary neutrons and delivered to the material surrounding the reaction vessel, where it becomes available as heat, which through heat exchangers may be used to generate steam and thereby power turbines and generators. The neutrons themselves may be used either to produce tritium from lithium or in a hybrid system to initiate fission reactions within a uranium blanket and in this manner further support the production of energy.

Two major strategies have been developed to solve the problem of plasma confinement at temperatures of 100 million degrees and more, both of which are still in an experimental stage. One of the two makes use of the principle of inertia, which is applied, for example, in the hydrogen bomb. In the initial phase of a fusion explosion, similar to any other explosion, it takes a certain though very short period of time to accelerate matter which still is sufficient for the release of a considerable amount of nuclear binding energy. In the hydrogen bomb the thermal fusion process is initiated by compression and heating of matter through the radiative pressure exerted by a fission explosion. The first hydrogen bomb in 1952 in the Enivetok Atoll demonstrated the feasibility of this conjecture but of course otherwise is a tragic and alarming document of misused scientific progress.

It is being tried to miniaturize this process for technical application in power stations. To this end pellets containing a fraction of a gram of a frozen deuterium-tritium mixture are dropped into a reaction vessel, where they are hit by either an intense laser or charged particle beam, thus being compressed and heated. In its final version the reaction vessel would probably have a diameter of some tens of meters and the energy release per explosion would be of the order of several 100 million Joule.

In another branch of fusion research efforts are made to confine the plasma with the help of magnetic fields. In the Tokomak-type installation the plasma is contained in a doughnut-shaped vessel in which a toroidal magnetic field is sustained by external electric currents within coils distributed around the plasma chamber. The installation is surrounded by a transformer yoke inducing a toroidal current inside the plasma. The plasma then is subject to constricting forces generated by this toroidal current.

The characteristics of the two approaches obviously are quite different, exhibiting high densities with short enclosure times in the first case, and low densities with comparatively long enclosure times in the second case. Power densities are high in inertial and low in magnetic confinement. Technical implications on

the way to construct a nuclear power station are extremely complicated in both cases.

9 Nuclear Fission

Some materials, which may be used as nuclear fuel, are listed at
the bottom right of Fig.4. Uranium-233, uranium-235, and pluto-
nium-239 especially are fissionable by thermal neutrons, the energy of which corresponds to the temperature of the material. Of
these isotopes only uranium-235 occurs in natural uranium, in our
times with a percentage of 0.7%, the rest is uranium-238. Uranium-
233 can be produced from thorium-232, plutonium-239 from uranium-
238 via neutron capture. Thorium-232 and uranium-238 are therefore referred to as fertile materials.

The fission process normally results in two fission products,
which incorporate quite different numbers of nucleons. They carry
away about 168 MeV, which is the major part of the total fission
energy of about 201 MeV. In most cases the fission products are
highly radioactive nuclei considered as waste.

Additionally, two or three secondary fast neutrons emerge from
the fission process with energies ranging between 1 and 2 MeV
typically. Over 99% of them are prompt neutrons released within
an extremely short period of time, not more then 1/100 of a millionth of a millionth second. The rest are delayed neutrons appearing with a mean delay time of about 10 s. These are essential for the control of power generation in nuclear reactors.

Uranium-233, uranium-235, and plutonium-239 are also fissionable
by fast neutrons. This is also the case — though with reduced
probability — for thorium-232 and uranium-238.

In order that a self-sustained chain reaction based on nuclear
fission can proceed, a sufficiently large amount of fissionable
material, called critical mass, must be put together to reduce
the surface-to-volume ratio and therefore the escape probability
of secondary neutrons, in order that at least one from each fission reaction should have a chance to initiate another fission
reaction.

If there are no or few nuclei of low atomic number present in the
material, the secondary neutrons will not lose much energy by
elastic scattering before inducing another fission process. The
chain reaction will then be sustained essentially by fast neutrons,
which is possible only in highly concentrated nuclear fuel. The
installation is then called a fast nuclear reactor.

An atomic bomb can also be considered as a fast reactor. If ura-
nium-235 is used as a nuclear explosive, it has to be enriched
to at least 90%. The critical mass of pure uranium-235 amounts
to about 15 kg, the one for plutonium-239 is about 4.5 kg. Still,
there are several essential differences between fast reactors
and atomic bombs. In the latter the chain reaction is sustained
essentially by prompt neutrons, while in the former the delayed

neutrons still contribute essentially. Furthermore, in an atomic bomb the critical condition is initiated by a chemical explosion compressing the fuel in order to have a very large release of energy in an extremely short period of time. Again, the principle of inertial containment is relevant. Since these conditions are not present in a fast power reactor, the latter cannot lead to an explosive release of energy comparable to an atomic bomb.

Under favorable conditions neutron economy in fast reactors allows to produce more plutonium-239 by breeding from uranium-238 than is consumed by fission as nuclear fuel. Therefore the development and use of fast power breeder reactors is a necessity if the major part of natural uranium, that is uranium-238, is intended to be made available as a source of energy.

A fast reactor power station producing 6000 million Watt of heat energy will typically contain a charge of 6000 kg plutonium as fuel and 100,000 kg uranium as fertile material. Liquid metal cooling is applied to prevent thermalization of neutrons.

Unfortunately, up to now, there is no more elegant way to transform nuclear energy into electricity than by the generation of heat, the production of steam, and the use of turbines and electric generators. As a consequence of the second law of thermodynamics, only one-third of heat energy finally is obtained as electricity. The rest of the energy so far in most cases is given away as waste. It is therefore worthwhile to consider possibilities to use this waste energy for heating purposes, especially so since a comparatively large portion of the total energy consumption for example in the Federal Republic of Germany is used in the form of low-temperature heat energy.

If a large amount of nuclei of low atomic number is present in the reactor, most secondary neutrons become thermalized by elastic scattering before they have a chance to induce another fission process and thus sustain the chain reaction. The installation is then called a thermal nuclear reactor.

If water is used as a moderator for the thermalization of neutrons and at the same time as a coolant to extract the heat energy from the reactor core, together with uranium as nuclear fuel the uranium-235 concentration must be enriched to more than 1%. A thermal reactor power station producing 3500 million Watt of heat energy will, for example, typically contain 100,000 kg uranium enriched to 3% uranium-235 as fuel.

Since the half-life of uranium-235 is 700 million years and therefore much shorter than the half-life of uranium-238, which is about 3.5 billion years, the uranium-235 content in natural uranium was much higher in earlier geological eras. The critical condition therefore could be established by nature in a rich vein of uranium ore embedded in an aquatic surrounding. This obviously happened in a place now called Okto in the Republic of Gabun, West Africa, 2 billion years ago, in the precambrian ages, as was disclosed by a routine analysis of uranium isotopes.

10   The Three Faces of Energy

In common language mass means as much as amount of matter. In physics, the notion of mass has two seemingly quite different meanings.

Inertial mass is a measure of the resistance of bodies against changes of their velocity. For example, the heavier a car is loaded, the larger is the force of breaking needed to bring it to a standstill within a certain interval of time. The special theory of relativity tells us that inertial mass and energy are actually two expressions of the same thing, the two quantities being related to each other through the famous formula $E = mc^2$. Since the velocity of light c, which appears in this relation, is a very large quantity, in terms of usual units, a small amount of mass corresponds to a very large amount of energy. 1 gram is equivalent to 90 million million Joule (Fig.6).

In chemical and even in nuclear reactions only a tiny fraction of the mass involved reappears as binding energy released in the form of kinetic or radiative energy. Even through, within the sun, nuclear binding energy released through fission reactions is equivalent to 4 million million grams per second.

A complete transformation of mass into radiative energy is possible only when certain additional conservation laws are fulfilled. In a far distant future, when man has learned to handle these conditions, matter and antimatter may eventually be used as a fuel to power space ships.

THE THREE FACES OF ENERGY

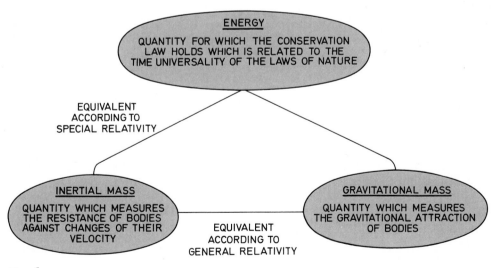

Fig.6

Heavy mass is a measure for the gravitational force acting between bodies. It has probably been noticed by Newton that inertial
mass and heavy mass are proportional to each other. But he certainly could not understand this fact. The equivalence of inertial
and heavy mass has become an ingredient of Einstein's general
theory of relativity. Internal energy, inertial mass, and heavy
mass of bodies are just different expressions for the same physical quantity.

The distribution of mass, more generally speaking energy momentum,
governs the geometric structure of space and time. It also determines the development of our universe as a whole, finite though
unlimited, whether it will expand for times to come or else finally collapse.

11 Time and Energy

At the very beginning of this treatise, we considered a swing as
an example for a physical system composed of two different objects, the earth and a human body bound together by gravitational
interaction. As mentioned before, the total energy of the system,
disregarding the kinetic energy of the earth, which is not relevant in this context, depends on the velocity and the height
above ground. The mathematical expression establishing this relation, called the Hamiltonian of the system, not only allows to
calculate the numerical value of the energy, but also through its
formal structure determines the change with time of the two dynamical variables mentioned above. In general, the energy function of any physical system governs its development in time. The
fundamental role of the Hamiltonian is demonstrated by the fact
that it may be understood as the formal equivalent of a physical
system.

Human intelligence in the course of biological evolution did not
have direct access to phenomena in the very large or in the very
small. This is why the laws of nature governing these two ranges,
the world of the universe on the one side, and of elementary particles on the other side of the scales, seem so strange to us.
The microcosmos is described by the formalism of quantum mechanics.
There, in the refinement of the theory, the energy function is
replaced by what is called the energy operator. But still the
latter determines the development with time of the physical systems.

As we try to focus our interest on very small objects like atoms,
nuclei, and elementary particles, we find that physical quantities
to some degree exhibit a fundamental lack of sharpness. Heisen-
berg's uncertainty relation states that energy and time can no
longer simultaneously be observed accurately. Virtual particles
may appear from the vacuum and live for extremely short periods
of time influencing the development of physical systems, thereby
leading us to attribute to them features of dualism and undeter-
minism. Even the notion of systems being composed or elementary
becomes questionable on this level of consideration.

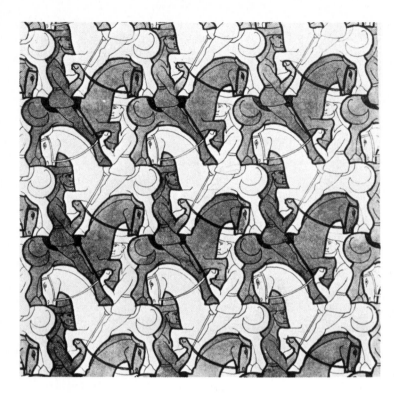

Fig.7

A strong heuristic principle guides us to search for the fundaments on which the huge and complex building of all scientific
knowledge rests. We have reason to believe that these fundaments
lie in the topological structure and symmetry properties of those
categories in which we formally may describe the laws of nature.
The universality of these laws expresses itself in their property
of being valid at all times, in the times when the earth was
formed from interplanetary dust, in our times, and still when man
has ceased to populate this planet. This is the principle of time
translation invariance. Let us illustrate this idea by an analogon
expressed in a fascinating piece of art by M.C. Escher, shown in
Fig.7. The picture reproduces itself when it is shifted by certain steps to the left or to the right. Similarly the laws of nature reproduce themselves when shifted to the past or future.
Noether's theorem states that each symmetry of this kind generates
a conservation law and hence a conserved quantity. The conserved
quantity originating from time translation invariance is energy.

12 Chaos and Structure

So far physical systems have been considered which consist of
just a few objects. A system consisting of many principally equal
objects, as the molecules of a gas confined in a vessel with iso-
lating walls, is shown schematically in the *upper part* of Fig.8.
As a consequence of the first law of thermodynamics the energy of
the gas does not change.

COMPLEX SYSTEMS

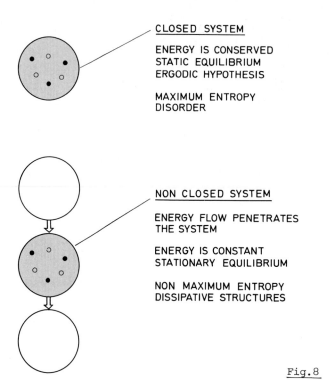

CLOSED SYSTEM

ENERGY IS CONSERVED
STATIC EQUILIBRIUM
ERGODIC HYPOTHESIS

MAXIMUM ENTROPY
DISORDER

NON CLOSED SYSTEM

ENERGY FLOW PENETRATES
THE SYSTEM

ENERGY IS CONSTANT
STATIONARY EQUILIBRIUM

NON MAXIMUM ENTROPY
DISSIPATIVE STRUCTURES

Fig.8

In order to understand its behavior one has to consider still another quantity, called entropy. Its development in time is determined by the second law of thermodynamics which in this context
appears in a somewhat different version. The entropy of an isolated system increases continuously and eventually approaches a
maximum value when the equilibrium has settled.

This feature is quite plausible, since one knows from statistical
mechanics that entropy is a measure of disorder. It appears to be
quite natural that by itself disorder is increasing with time in
any closed system. The dynamical variables of objects in the equilibrium follow a statistical distribution, which as a consequence
of the ergodic hypothesis, is determined by the energy function.
The Hamiltonian thus governs the state of equilibrium in systems
composed of large numbers of objects.

If one adopts the validity of the second law of thermodynamics
also for the universe as a closed physical system, the latter
consequently develops into a state of maximum entropy, a uniform
chaos.

In the middle of the *lower part* of Fig.8 again the same physical
system is shown in a different situation. It is no longer isolated since there are two other systems, shown above and below

respectively, one of which feeds energy into, while the other at
the same time rate extracts energy from the system under consideration.

Still the total energy is conserved though it is not a closed
system. The second law of thermodynamics does not hold. Entropy
and disorder need not increase continuously. Under these conditions of penetration by a flow of energy dissipative structures
may arise within the system, as has been shown by Prigogine, that
is an orderly behavior of matter in space and time.

Many examples of dissipative structures are known from chemistry
and physics, for example the appearance of convective zones within some layers of the sun. The discovery of dissipative structures may also deliver the key to understand the origin of organic
life, which may turn out to be a functional dissipative structure.
But this still remains speculative.

13   Demands and Resources

Economic systems like the Federal Republic of Germany are also
penetrated by a flow of energy. May such economic and social
structures in a more general sense be understood as functional
dissipative structures?

The relation between the per capita consumption of energy and the
gross social product illustrates this idea, since it does not represent only the momentary ensemble of systems but also the development of individual systems in the course of time.

The need for energy in the world will be accelerated dramatically
not only by the increase of world population but also by economic
development of the still not industrialized countries.

For how long can these needs be met by fossil fuels, do we really
need nuclear energy? The existing economic and political structures let us ask this question more specifically. Where and at
what price sources of primary energy are available, and how is
the demand constituted of different forms of energy, electricity,
heat, and fuel?

For many decades or even hundreds of years coal can cover the major part of energy demand where it is available at low prices and
where no environmental problems arise. As it seems, the United
States could do well without nuclear energy far into the next century. This seems not to be the case for the Federal Republic of
Germany and even less for the Republic of France.

The overproportional increase of oil consumption has to be confronted with the prognosis that the production will culminate
within one generation's time and then decrease. Therefore, sub-
stitutes for natural petroleum to power motor vehicles have to
be developed still within the time of our generation. The most
promising candidate for the near future is certainly the production of synthetic fuel from coal, especially in those regions
where the latter is abundant and low priced.

Still, we have to think of other options to convert primary energy from coal, from the sun, and from nuclear fuel into forms appropriate to be stored safely and compactly in vehicles. Batteries combined with electric motors offer one such possibility, the major handicap being that the net energy density per unit of weight at the present level of technology is small, about 1% for lead-acid batteries and about 3% for lithium-metal-sulfide batteries, as compared with gasoline.

The same disadvantage is present in the combination of hydrogen as fuel stored in the form of metal hydrides with combustion engines, where, e.g., in the case of iron-titanium-hydride the energy density per unit of weight is about 5% of that for gasoline. Both options, electricity as well as hydrogen, offer the great advantage of being virtually non-polluting.

Alternative forms of energy, like solar radiation, can cover a certain fraction of the total energy demand in regions with favorable climatic, meteorological, or geological conditions. In the Federal Republic of Germany this fraction will certainly not exceed a few percent.

The power density of solar radiation outside the atmosphere, the solar constant, is 1353 J per second and square meter, of which about 34%, the albedo, are scattered back into space by the earth's atmosphere and surface. Still about 17% are absorbed within the atmosphere. Actually, the low power density of solar radiation is a major problem restricting the possible construction of power stations to desert zones where, for example, hydrogen could be produced from water for transportation to more industrialized regions. In view of the fact that the total global solar radiation absorbed by the earth's surface amounts to about 3000 billion billion Joule per year and thus exceeds by about 10,000 times the total world energy consumption, solar radiation may become in the long term a globally important alternative form of energy.

14 Problems Still to Be Solved

Public discussion on the risks of nuclear technology primarily refers to the problem that radioactive material must not escape from nuclear power stations. The technical possibilities to meet this requirement seem to be given. Still, the technical development is not yet concluded. Should nuclear reactors be built deep underground in the densely populated regions of the world, especially in the weather zones of world politics, as in Central Europe?

Large amounts of radioactive fission products are deposited today for example in geologically stable salt stocks to ensure that they cannot escape control. Is it necessary to keep this waste manipulatable and hand it over from generation to generation?

The development of nuclear technology inevitably leads to the proliferation of nuclear know-how and the possession of highly concentrated fuel. Like an avalanche, the danger of military misuse spreads over the world. Will strategies be found and world-wide mechanisms be installed to ensure that this development does not lead to self-destruction?

These problems have to be solved. A new dimension of responsibility for life on earth and for the generations to come has been put onto mankind. Still, the energy problem is related to and embedded in other problems, such as maintaining the economic and political balance and stability in this world. The never ending, always accelerating scientific and technological progress brings about new opportunities and new risks, and the time scale of evolution in which man has to adjust himself to the pace of progress contracts.

Resources and Reserves of Fossil and Nuclear Fuels

F. Bender and K. E. Koch[1]

1 Introduction

The drastic price hike for petroleum — ca. 1500% since 1960 — has
made it clear to everyone that there are limits to the availability of this source of energy. Therefore, efforts have been increased in the search for and utilization of other sources of energy. It has been, and still is, necessary to develop new processes
to guarantee the supply of energy, the basis of economic stability and growth, for the medium- and long-term future. Starting
points of such considerations are the following questions: What
quantities of non-renewable raw materials exist? How long will
they last? For this reason global surveys of reserves and resources of energy raw materials have been made. The number of
such surveys has increased during the past several years. The
results of these surveys may differ in details, but the overall
picture is of the same order of magnitude. The limits are visible
and thus the limits of growth as well. Nevertheless, these limits
are not menacingly near, and they can be pushed even further into
the future, if the necessary efforts are made.

What are the limits for the different energy resources at present?
What can and what must be done to push them back?

2 Coal

The term coal encompasses solid fossil fuels with a wide range
of energy content. Depending on the degree of coalification, this
is expressed in tonnes of coal equivalent (1 tce = 1.5×10^6 kcal).
One metric tonne of

lignite corresponds to	$0.3 - 0.6$	tce,
subbituminous coal corresponds to	0.78	tce,
bituminous coal and anthracite corresponds to	1	tce.

Due to the increase in the price of oil, many countries have begun to reevaluate their resources of coal. They have started exploration campaigns in areas around known coal occurrences and

1 Bundesanstalt für Geowissenschaften und Rohstoffe, Postfach 51 o1 53,
  3000 Hannover, FRG

in prospective areas. Programs for the development of new processes and technologies for coal conversion have been started for
surface installations and for coal in situ.

The regional distribution of coal, as well as the prospective
areas, are basically known. New discoveries of coal occurrences
the size of the Ruhr District in the Federal Republic of Germany
or of the Appalachian basin in North America can hardly be expected any more. Nevertheless, the possibility of finding numerous smaller occurrences does exist, especially of lignite with a
volume of several 100 million to a few billion tonnes.

All quantitative estimates show that coal is the most abundant
and widely distributed primary energy resource. On the basis of
present production, measured reserves of hard coal have a static
lifetime of about 100 years, those of lignite of about 130 years.

The energy content of the proven recoverable coal reserves (about
688×10^9 tce) is about 1.8 times larger than the corresponding
reserve category for all hydrocarbons (petroleum, natural gas,
bituminous shale and tar sands). The corresponding ratio is even
greater for resources, ca. $10,690 \times 10^9$ tce or 5.8 times.

The regional distribution of bituminous coal and anthracite
(Fig.1), and lignite (Fig.2), shows the great difference between

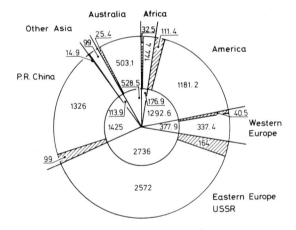

BITUMINOUS COAL
AND ANTHRAZITE

Proved recoverable
reserves and additio-
nal resources in situ
in 10^9 t CE

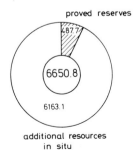

Fig.1

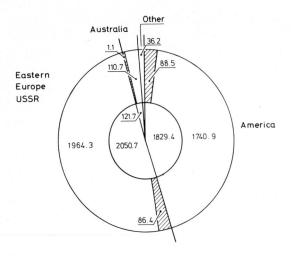

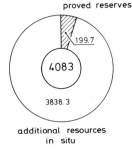

SUBBITUMINOUS
COAL AND LIGNITE

Proved recoverable reserves and additional resources in situ
in 10^9 tCE

Fig.2

proven recoverable reserves and additional in-situ resources. This large difference makes it quite evident that one of the main tasks of the future will be to transform as much of the resources as possible into recoverable reserves.

The development of new technologies for coal processing (e.g., gasification and liquification) is an important step in this direction. At present, operations of this kind on production scale are installed at the surface; underground installations are being planned or are on pilot scale. The biggest complex in operation in South Africa produces with a capacity approaching approx. 5 million t of "synfuel" per year.

3   Petroleum

The prerequisites for an oil deposit are the presence of a sedimentary basin with source rocks, porous and permeable reservoir rocks, impermeable cover rocks, and the possibility of accumulation in traps. Most of the known deposits are found in anticlines. Others occur in stratigraphical or other structural traps.

About 600 sedimentary basins are known which offer favorable conditions for the formation and accumulation of liquid hydrocarbons.

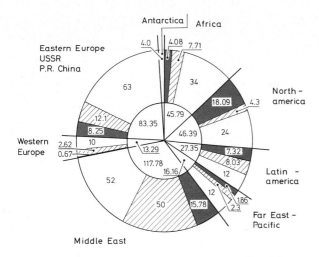

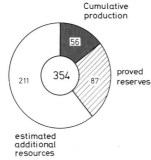

PETROLEUM

Cumulative production (1.1.1980), proved recover-able reserves and additional recoverable resources in 10^9 t

Fig.3

In 160 of these basins, exploration has led to the discovery of oil fields known today. A further 250 basins have been partially explored without finding commercially exploitable deposits. About 200 basins are practically unexplored. Most of them are located in hostile environments (remote continental areas, arctic regions, offshore areas of more than 200 m water depth).

The prospective areas for hydrocarbons total about 74 million km^2 onshore, about 23 million km^2 offshore. The latter is understood to comprise the shelf region as far as the continental margin.

Most of the more recent estimates of total recoverable petroleum (cumulative production + measured recoverable reserves + estimated additional recoverable resources) range between 250×10^9 and 360×10^9 t. The value of 354×10^9 t given here is within this range and can be regarded as a conservative estimate.

Figure 3 shows how cumulative production, measured reserves, and estimated additionally recoverable resources are distributed throughout the world.

The Middle East will remain the most important producing region for a long time and will have a considerable resources potential.

Countries with centrally-planned economies — especially the U.S.S.R. and the People's Republic of China — as a whole have as well a considerable potential. The relatively recent development of exploration and exploitation is indicated by the low value for cumulative production and high values for measured reserves (Fig.3).

For North America, a different picture results: There, the cumulative production amounts already to about 40% of the total resources, whereas measured reserves amount to less than 20% of the resources. This is a situation which applies to most of the industrialized western countries: a long history of exploitation≈, high cumulative production, ≈high proportion of oil fields nearing depletion.

This situation has stimulated the development of enhanced recovery methods to improve the recovery factor, which at present averages worldwide about 35% only, in many cases considerably lower.

It is difficult to estimate how far the recovery factor can be improved. Under favorable conditions it may be expected that an additional 50% over primary recovery may be gained.

At present the static lifetime of measured reserves of oil is about 28 years.

Every estimate of reserves can only represent a momentary picture based on current knowledge. It will be quite possible to prolong this lifetime by new discoveries and by enhanced recovery methods; an end of the age of oil, however, can be seen on the horizon.

4 Natural Gas

According to their origin, the following types of natural gas can be distinguished:

> gas formed under the same or similar conditions as petroleum;
> gas formed during the coalification of coal or of dispersed fossilized plant material;
> gas from bituminous shale;
> gas formed by bacterial action.

In principle, migration and accumulation of gaseous hydrocarbons are governed by the same parameters as for petroleum. Because of the considerably smaller size of gaseous hydrocarbon molecules compared to liquid hydrocarbons, gas can more easily pass through porous rocks and can accumulate in denser reservoir rocks.

The measured recoverable reserves of natural gas amount to more than 70×10^{12} m^3. While measured reserves of oil have changed little during the last 5 years, the corresponding figures for natural gas have shown a pronounced upward trend. This is mainly due to the development of the necessary distribution systems: pipelines, tankers for Liquid Natural Gas (LNG). Discoveries of natural gas, formerly rather regarded as a "mishap" in oil exploration, today are the result of targeted exploration activity.

The difficulties and problems connected with the exploitation of natural gas may be demonstrable by the largest gas field of the U.S.S.R., the Urengoiskoye field. They result mainly from the construction of the distributing system under difficult infra-structural conditions. In the mid-80s, it is expected that pro-duction from this field will reach 200-250 × 10^9 m^3 per year.

Varying amounts of gas are produced together with petroleum (as-sociated gas). Often this gas is flared for the lack of a distri-buting system, especially in the Near East. In the U.S.S.R. too, natural gas in the order of several 10 × 10^9 m^3/year is flared because in recently developed oil provinces (e.g., Tyumen in Western Siberia or Emba at the Caspian Sea) the construction of installations for separation and distribution of gas has not kept place with oil production.

Production in the two most important producing regions, North America and U.S.S.R./China/Eastern Europe, showed clearly oppo-site trends between 1972 and 1979: During the same period in which production in North America decreased by about 100 × 10^9 m^3 to 628 × 10^9 m^3, production in countries with centrally-planned economies doubled to 512 × 10^9 m^3. Aside from other factors, this is a result of the restrictive price policy in the USA, which has had a hampering effect on development.

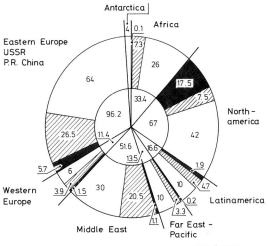

NATURAL GAS

Cumulative production, proved recoverable reserves, estimated additional recoverable resources and total resources in 10^{12} Nm^3

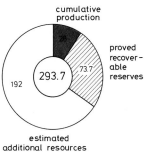

Fig.4

The future general development of natural gas production is difficult to predict (Fig.4). Nevertheless, it must be more optimistically assessed than for oil. It may be expected that total annual production of natural gas will reach a maximum of about 2200×10^9 m^3 around the year 2000.

On the basis of present production, the static lifetime of measured reserves of gas is about 45 years.

It can be assumed that the price for natural gas will be increasingly oriented to the price of petroleum. This adjustment of price (1 toe corresponds roughly to 1150 Nm^3 of saleable gas) will stimulate the exploration. This will help upgrade a large portion of additional resources into recoverable reserves.

5 Bituminous Shale and Sand

Argillaceous and/or calcareous sediments with more than 40 l of recoverable hydrocarbons per tonne of rock are understood under the term bituminous shale.

Bituminous sands also contain an elevated proportion of hydrocarbons. The "Alberta Sands" (Canada), for example, average 83% of sand, 13% of bituminous matter and 4% of water.

The heavy oils from shale and sand are characterized by an unfavorable hydrogen/carbon ratio. Therefore, hydrogen has to be added during upgrading refining.

The large deposits of bituminous shale in North America have been known for a long time. They represent more than 65% of the proven world reserves (Fig.5). The thickness of the bituminous shale there reaches more than 300 m in some places. Recovery of oil averages 110 l/t of shale.

At present, a recovery of 40 l/t of shale is regarded as the lower limit of profitability.

Today, major production of oil from shale takes place only in the U.S.S.R. (about 40 million t/year) and in the People's Republic of China (about 6 million t/year).

In the United States, considerable efforts are being made to develop processes to exploit the considerable reserves. It seems probable that large-scale production will begin before the end of the 1980s.

Large deposits of bituminous sands are known in North and South America. The largest, the "Orinoco Tar Sands" in Venezuela, stretches in one direction for about 400 km. Another important field is that of the "Athabasca Tar Sands" in Canada, which cover an area of about 50,000 km^2. The first experimental installations

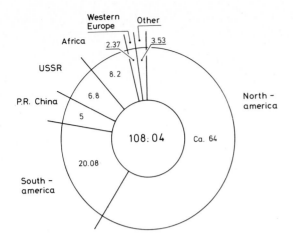

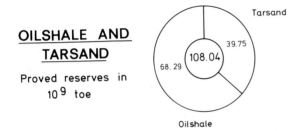

OILSHALE AND TARSAND

Proved reserves in 10^9 toe

Fig.5

there have been producing 2.5 million t of oil per year since the 60s. Others took up operations in the late 70s, and are aiming at an oil production of 6.5 million t per year.

These activities are confined to opencut operations. In-situ methods are still in the pilot stage.

On the basis of even relatively recent surveys, the impression could result that deposits of so-called unconventional hydrocarbons from bituminous shale and sand are restricted to only a few regions (Fig.5). But it has to be considered that in the past petroleum was easily and cheaply available. Therefore, a systematic, worldwide exploration for bituminous shale and sand has not taken place. Even the reserves of some known deposits have been only partially determined.

The data on reserves and additional resources are so incomplete that in Fig.5 the additional resources have been neglected and only the measured reserves of oil from shale and sands are shown.

Nevertheless, the oil content of the measured reserves amounts to about 108×10^9 t, which is already 2.5 times the value of the corresponding category for conventional petroleum. The factor

for the estimated additional resources of oil from shale and sand
(416×10^9 toe) is almost 2.

From these figures it becomes evident that considerable efforts
should and must be made in the future to systematically explore
the considerable potential of bituminous shale and sand, and to
develop technologies for their utilization.

6 Uranium and Thorium

Delays in construction of nuclear power plants have led to a decrease in exploration for uranium. This tendency is understandable if one regards the development of the price for natural uranium ($/lb U_3O_8) during the last 3 years (Fig.6). The rapid de-
cline in price from about $ 45 to less than $ 30 makes costly
prospecting and exploration seem little rewarding. Even some producing mines are considering closing down operations because of
this development.

This has led to a situation in which (Fig.7) measured uranium
reserves in some regions are considerably greater than estimated
additional reserves, for instance in Western Europe, Africa, and
especially in Australia. Mining companies show little inclination
to explore for more reserves than are sufficient for the present.

A similar picture can be drawn for thorium. Known resources have
a lifetime reaching to the year 2025. The picture of the reserves/
resources situation is so rudimentary that no graph is given here
for thorium.

The regional distribution of the resources for uranium (Fig.7)
shows that a considerable potential is still present. The measured reserves (about 3.053×10^6 t U) amount to almost 75% of
the estimated reserves (4.094×10^6 t U), but to only 10% of the
speculative reserves (about 29.292×10^6 t U).

Even if the reserves of nuclear fuels today only represent about
25% of the total reserves of energy resources, it has to be considered that this quantitative comparison of energy only takes
into account the U^{235} in natural uranium (7 kg U^{235}/t U_{nat}) used
in conventional reactors in an enriched form.

For nuclear fuel, production capacity is a much more important
criterion for the supply situation than the quantity of reserves.
The reason for this is the complex technology from exploitation
to generation of electricity, as well as reprocessing and final
waste disposal.

It seems very unlikely that the reserves of uranium will last
only about 50 years, as some forecasts predict. On the other hand,
present technology for the use of nuclear fuel does not represent
the last word. Advanced technologies which make use of thorium,
as well as plutonium and U^{233} from breeder reactors, would better

Price Range for Immediate Deliveries of Natural Uranium – Europe

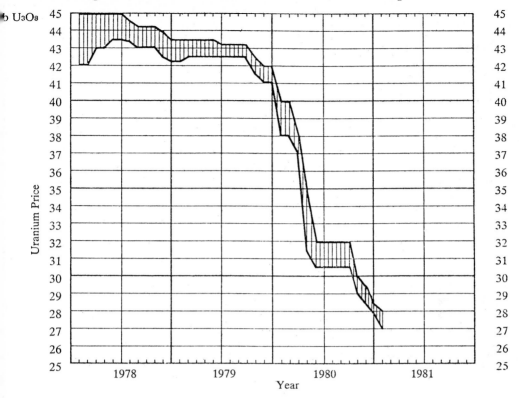

Fig.6   NUKEM Market Report   1/81

utilize the energy potential of these resources. It is too early
to include nuclear fusion in energy considerations.

A comparison of energy production from U^{235} in different types of
reactors makes evident the development that is already possible.

If the quantity of energy from the normally used light-water reactors is set at 1, the energy that could be obtained from a high-
temperature reactor is 7.5 and about 60 from a fast-breeder reactor.

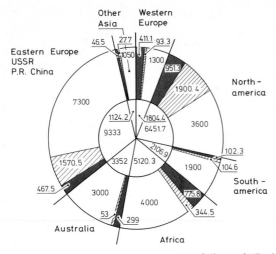

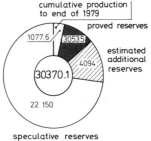

URANIUM

Proved, estimated
additional and
speculative reserves
in 10^3 t U, price -
cathegory up to
130 $ / kg U

Fig.7

7 Future Trends and Consequences

It would be wrong to draw too black a picture of the future energy supply, as has been done in some studies. Nevertheless,
these pessimistic views have created an energy consciousness.
But crisis-like difficulties in the energy supply had to arise
before efforts were accelerated to safeguard the supply:

the drastic rise in the price of petroleum,

the uneven regional distribution of energy resources,

the political and economic opportunities resulting from these
facts for those countries possessing resources,

the increasing political and economic risks for those countries depending on imports.

The increasing opportunities for the producing countries, and
the increasing risks for the consumer countries have contributed
to the fact that the price of energy raw materials, especially
hydrocarbons, is regulated less and less by supply and demand.
To neglect this trend instead of including it in the considerations would have serious consequences.

What are the consequences to be drawn from the reserves situation for energy resources?

Regarding the period until the beginning of the next decade, the absolute consumption of petroleum and natural gas can still be slightly increased, but relative consumption must be significantly reduced by conservation and substitution. The increased use of coal and nuclear energy is possible and must be promoted accordingly. Production of petroleum will have peaked by the year 2000; that of natural gas only a little later. Most of the increase in energy consumption must be filled by coal and nuclear energy. Additionally, greater utilization of bituminous shale and sand will be necessary.

After the year 2000, petroleum and natural gas will be used primarily by the petrochemical industry. Production of oil from shale and sand will be increased further. Energy demand will have to be covered mainly by coal and nuclear energy. Part of the energy supply will also come from alternative energy sources, mainly from the unlimited reservoir of geothermal and solar energy.

Nature has not yet set the time for the depletion of energy resources in the near future. The problems of energy supply, therefore, are solvable on the condition that

> intensive programs are launched for exploring for energy resources and transforming them into economically exploitable reserves,

> efforts for the technological development of the use of energy resources, primarily coal and nuclear fuels, is intensified to achieve an optimum degree of utilization,

> economical and political conditions are established to guarantee that all possibilities are used steadily and speedily.

References

Bender F (1979) Sufficient energy raw materials for everyone. Proc Vol Int Centennial Symp "Resources for the twenty-first century". USGS, Reston, Va, pp 14-19

Bender F (1980) Survey of energy resources. Erdöl und Kohle, 11. Weltenergie-konferenz München zur Präsentation des unter der Gesamtverantwortung des Consultative Panel der WEK von der Bundesanstalt für Geowissenschaften und Rohstoffe, Hannover, erstellten "Survey of Energy Resources 1980"

Bonse W (1980) Möglichkeiten der Ölgewinnung aus Schweröl- und Bitumenvorkommen. Erdöl Erdgas Z 96/12

World Energy Conference (1980) Survey of energy resources 1980. Prepared by federal inst Geosci Nat Res (BGR), Hannover, Sept 1980. Koch J, Solid fossil fuels. Schubert E, Hydrocarbons. Meyer J, Nuclear fuels

Synthetic Fuels[1]

W. Peters[2]

1 The Role of Coal Processing for the Supply of Energy in the Future

The availability of natural gas and crude oil can, in the future,
no longer keep up with the increasing demand. There will inevitably be a global shortage of crude oil supplies in the 1980s
and of natural gas supplies in the 1990s. This, now widespread
realization, that was slowly growing discernible approximately
10 years ago, has triggered off energetic reactions.

> Motivated by irrelevant political affairs, the oil producing
> countries have been staging temporary oil supply crises since
> 1973 in order to use them as an excuse for phantastic price
> increases, long before there is really a shortage of crude
> oil.

The oil consuming countries, especially the industrialized countries have, however, still only too half-heartedly started to
reduce energy consumption through more efficient application and
by abandoning their consumer habits. Furthermore, remarkable
initiatives are being taken to develop alternative ways of cop-
ing with the shortage of supplies which is, nevertheless, to be
expected.

> The alternative solutions being discussed are manifold: Util-
> ization of solar radiation, wind, geothermal heat, tidal force,
> biomass, etc., and especially the increased employment of coal
> and nuclear energy. While the kinds of energy first mentioned
> will gain importance regionally but not globally, the main
> substitutes for mineral oil and natural gas will be nuclear
> energy and coal.

The already existing consumer structures will play an important
role in the substitution of mineral oil and natural gas by coal.
The Federal Republic of Germany, for example, disposes of a widely
spread gas supply network with a total length of more than
120,000 km. This supply system and the hope of its continued use
even after national exploitation and imports of natural gas are
no longer sufficient to satisfy the overall demand, is one in-

---

1 This paper was translated by Mrs. S. Messele-Wieser

2 Steinkohlenbergbauverein, Franz-Fischer-Weg 61, 4300 Essen 12 (Kray), FRG

centive for producing gas from coal. The petroleum branch, too, disposes of a supply system (set up with high investment costs) consisting of oil pipelines, tankers, etc., which could also motivate the production of oil from coal, once oil imports are reduced. In addition, coal will be used for direct heat production and for electricity generation for the medium-, and weak-load range.

Taking these circumstances into account, there is clearly a competitive situation between the many different conversion processes of coal into forms of secondary energy. The main task then is to pick out the optimal process based on the criteria of global efficiency, ecological beneficence and profitability. In the following part, the main conversion techniques will be roughly explained, so that in the end the first outlines of a research and development strategy can be devised.

2 Gas from Coal

During gasification (Fig.1, *right*) coal is cracked into smallest combustible gas molecules of hydrogen and carbon monoxide at a temperature above 900°C and with the help of steam, that decomposes during the process. A considerable heat input is needed for gasification: this can be achieved by adding oxygen and by combustion of one-third of the coal in the gasification reactor.

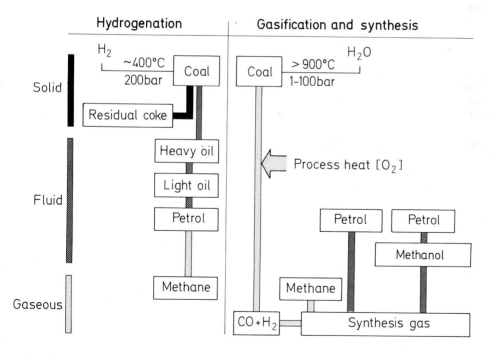

Fig. 1. Methods for the production of fluid and gaseous products from coal

34

This is called an autothermal process. Another possibility would
be to take in the process heat of more than 900°C from a high-
temperature reactor, by means of a heat exchanger. This is then
called an allothermal process.

The resulting gas mixture of hydrogen and carbon monoxide could
already be used as fuel. It is, however, more advantageous to
apply it as a synthesis gas for the catalytic synthesis of hydrocarbons: in this way the methanation leads to substitute natural
gas (SNG), the Fischer-Tropsch Synthesis to chemical products or
petrol, the methanol synthesis to methanol, and finally the Mobil
process-reaction leads to petrol. In addition, as final product
hydrogen can be extracted through shift-conversion. In the future, hydrogen is expected to be used as fuel.

At the moment more than 35 process configurations exist for the
gasification of coal, however only few of them have been developed for large-scale technology. In Fig.2, three basic types of
the already commercially operating autothermal coal gasification
processes will be compared with each other.

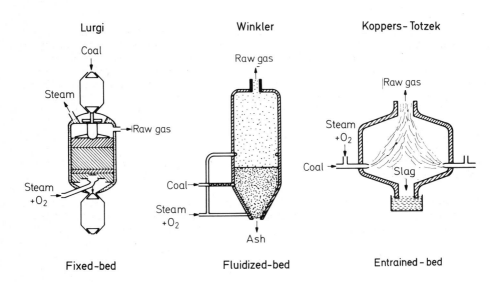

Fig. 2. Types of industrial gas generators

The fixed-bed reactor is represented by the Lurgi gasifier
(Fig.2, *left*) that has proved successful for decades. Here the
coal is fed to the reactor through a lock hopper system at the
top and passes through the generator from the top downward. The
gasification medium, steam and oxygen is introduced from the
bottom and flows counter current to the coal.

The result is a mixture consisting mainly of hydrogen, carbon
monoxide, and a considerable amount of methane. Because of the
counterflow principle this process is the most profitable as far
as the thermal efficiency is concerned. It has, however, the

disadvantage that it cannot operate with every kind of coal, so
that before being utilized in the gasifier fine coal for instance
has to be first agglomerated. All excessively caking coals cause
the coal pile in the generator to sag. The main development task
is an increase of output by raising the gasification pressure to
100 bar and the gasification temperature above the ash melting
point.

The second processing system is represented by the fluidized-bed,
as developed by Winkler in the 20s (Fig.2, *center*). The cocurrent
flow principle here operates at a medium temperature that must
be below the ash melting point of coal. Considering the energy
output, this process stands between the fixed-bed reactor and
the entrained-bed reactor.

One disadvantage is the fact that fuels with a low reaction velocity cannot be utilized. Actually, only lignite can be gasified
with a satisfactory conversion rate at atmospheric pressure.
Caking coals, as mainly found in the Ruhr Region, cause difficulties in the fluidized-bed gasifier. The further development
of this process aims at a higher operating temperature and higher
operating pressures.

The third processing scheme is represented by the entrained-bed
gasifier (Fig.2, *right*, schematic representation of the Koppers-
Totzek process). Here, too, the gasification medium flows cocur-
rently with the coal. Conversion takes place at a peak temperature of 1500°C with a high reaction rate. The advantage of this
process is its independence of the type of coal applied, so that
all types, first finely crushed, can be used. Compared to the
ones mentioned above, this process has less thermal efficiency,
due to the cocurrent principle and the high gasification temperatures. Since the entrained-bed process has, up to now, been
operating without pressure, attempts are being made to adapt it
to higher gasification pressures.

Allothermal processes are considered an alternative to the above
mentioned autothermal processes. Here the reaction heat is generated outside the reactor e.g., in a high-temperature reactor,
and then transferred to the gasifier by direct heat exchange.

One concept for the processing variations being considered is
shown in Fig.3. The coupling of nuclear reactor and gasifier
takes place over a gas circuit, operating with helium as heat
transfer medium. In the nuclear reactor, helium is heated to
950°C through the heat generated by nuclear fission. The superheated helium in the primary circuit flows through an intermediate
heat exchanger. In a secondary loop another helium circuit is
operated, in which a heat register is inserted. Over its pipe
system inside a fluidized-bed gasifier heat is transferred to
the gasification bed. Thereby the helium cools down to a gasification temperature of 700°C with lignite, and of approximately
800°C with hard coal. The raw gas produced in the gasifier can
be processed into synthesis gas or SNG. The advantage of this
concept is the almost twofold output of gas per unit of coal
processed by the plant, whereas the CO_2 emissions are only half
as high.

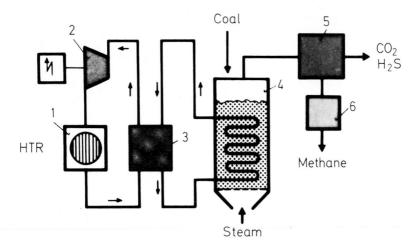

Fig. 3. Coal gasification with HTR-heat [immersion heater principle]. *1* High
temperature nuclear reactor; *2* helium turbine; *3* heat exchanger; *4* gasification reactor; *5* gas purification; *6* methanation

This process is at present being successfully operated in a
semitechnical gasification plant with an hourly capacity of
200 kg coal. Two further intermediate steps will have to be
taken before a large-scale technical plant can be constructed.
This process will hardly be commercially available before the
end of the century.

In Fig.4 the present activities in the field of gasification
in the Federal Republic of Germany are summarized, to which no
further details will be given. Nine firms (column 2) as well as
several manufacturers of plant equipments are participating in
the realization of the projects and have joined in associations.
At the moment the total costs for these programs are estimated
to amount to approximately DM 500 million by 1983. These expenses are being paid mainly by the "Bundesministerium für
Forschung und Technologie" (Federal Ministry for Research and
Technology), and also by the North Rhine Palatinate and the participating firms. The large-scale development program could be
concluded by 1983/84, so that by 1982 decisions could be made
as to which project should be further developed. Applications
for subsidies have been made to the Federal Government for the
construction of 11 gasification plants operating by the processes 1-7 as shown in Fig.4. Which project is to be realized will
be decided after submission of the preliminary projects.

This expensive large-scale testing program will be pushed through,
although gasification has not proved profitable yet. The possible
costs of a commercial plant based on the Lurgi gasification process with subsequent methanation, are classified in Table 1. For
example 2.25 billion m^3/y of substitute natural gas could be produced from 4.5 million t/y of coal. From the estimated investment
costs of roughly DM 3.2 billion and operating costs of DM 1.68
billion, the gas production costs are computed to be DM 0.75 per

No.	Process	Management	Location	Size of plant	Commencement of operation
1	Saarberg-Otto	Saarbergwerke, Dr. C. Otto	Völklingen	11t/h	1979
2	Ruhr 100 (Lurgi)	Ruhrgas, Ruhrkohle Steag	Dorsten	7t/h	1979
3	Texaco	Ruhrkohle, Ruhrchemie	Oberhausen	6t/h	1978
4	Shell-Koppers	Shell	Hamburg	6t/h	1979
5	KGN-process	Kohlegas Nordrhein	Hückelhoven	1t/h	1979
6	High temperature-Winkler	Rheinische Braun-kohlenwerke	Frechen	1t/h	1978
7	VEW coal conversion	Vereinigte Elektrizitätswerke	Werne	1t/h	1976
8	Steam gasification (with nuclear energy)	Bergbau-Forschung	Essen	0.2t/h	1976
9	Hydro-gasification (with nuclear energy)	Rheinische Braun-kohlenwerke	Wesseling	0.2t/h	1975

Fig. 4. Pilot plants for the gasification of coal in the Federal Republic of Germany

Table 1. Design data of a commercial gasification plant (April 1981)

Coal input	4.5 Mio t ce/y
Gas production	2.25 Billion m_n^3/y
By-products	0.1 Mio t/y
Investment costs	3200 Mio DM
Operation costs	1680 Mio DM/y
Production costs per m^3	75 Pf/m^3 (H_o)
Selling price	101 Pf/m^3

m^3. The consumer would have to reckon with a price of roughly DM 1.01 per m^3. This price may, at the moment, be twice as high as that of natural gas now being imported under a new delivery contract. Through utilization of nuclear gasification a reduction of the gas-producing costs by perhaps 20% is hoped for.

3 Liquid Products from Coal

The hydrogenation (Fig.1, *left*) at moderate temperatures of 480°C, constitutes a light pyrolysis in which the large carbon molecules gradually break up into smaller ones. At high pres-

sures above 200 bar the hydrogen saturates the released valen-
cies at the point of fracture. In this way low-molecular compounds with higher hydrogen contents develop, which subsequently
change into the fluid or gaseous aggregate. Depending upon the
catalysts added and the time allowed for reacting, it is possible
to produce heavy oil, light oil, or gasoline. By using higher
temperatures of approximately 800°C and by reducing the pressure
below 100 bar, the liquid phases can be skipped, so as to produce methane directly. This is also called hydro-gasification.
The disadvantage of hydrogenation is the fact that not all the
coal is converted, thus leaving residual coke that can, however,
be processed into hydrogen in a gasification plant or, into
steam or electricity in a steam boiler.

It is well known that from 1913 on, Bergius already carried out
the first hydrogenating experiments in autoclaves, which were
followed by further experiments in small-scale technical plants.
Subsequently other tests, done by the IG-Farbenindustrie, led
to the basis for a technical process which facilitated the construction of the first large-scale research plant in Leuna, 50
years ago. Up to 1945 a total of 12 hydrogenation plants were
constructed in Germany reaching a capacity of almost 4 million
tons/y during 1943/44. After the war the victorious powers
ordered the shutting down of these plants. Later, owing to cheap
oil products, coal hydrogenation was no longer interesting from
the economic point of view.

After the 1973/74 oil crisis it was necessary to start from the
beginning, since the methods of hydrogenation had, in the meantime, been forgotten. For this reason two technical plants were
set up at the Saarbergwerke AG, Saarbrücken, and at the Bergbau-
Forschung GmbH, Essen (Fig.5) on which no further details can be
given here. The experimental operation of several years has, in
the meantime, produced important construction data and operational developments compared to earlier techniques.

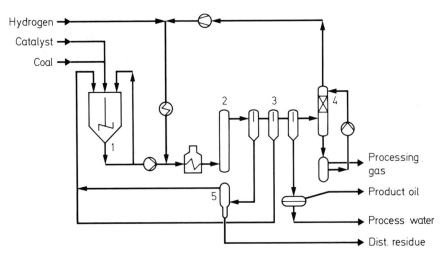

Fig. 5. Experimental plant for coal liquefaction. 1 Slurry preparation;
2 conversion reactor; 3 separation; 4 recycle gas scrubbing; 5 vacuum flash

In the next, they are to be technically tested in two large
testing plants with a coal capacity of 200 t/d at the Ruhrkohle
AG and VEBA Oel AG, and of 6 t/d at the Saarbergwerke AG. Both
testing plants are in operation since 1981.

The development of coal hydrogenation in the Federal Republic of
Germany may cost more than 300 million DM by 1983, expenses will
be met mainly by the North Rhine Palatinate, the Federal Ministry
for Research and Technology, as well as by the participating
firms. This is the second focal point of the research program
under way in the Federal Republic of Germany. In addition, German firms, especially the Ruhrkohle AG supported by the Federal
Ministry for Research and Technology, are participating in processing developments in the United States.

Although the described program is not yet concluded, the construction of large liquefaction plants is now being considered.
Three applications for subsidies have been made to the Federal
Government. A production plant as schematized in Table 2 could
be available by the end of the 80s.

Table 2. Design data of a commercial liquefaction plant
(April 1981)

Coal input	6 Mio t ce/y	
Primary products		
– Light destillates	1.64	Mio t/y
– C_2-C_4-gases	0.36	Mio t/y
By-products	0.14	Mio t/y
Investment costs	4400	Mio DM
Operation costs per year	2480	Mio DM
Product costs	96	Pf/l
Selling price	182	Pf/l

In this plant, roughly 1.6 million tons of light distillates
that can be processed into petrol, and roughly 0.4 million tons
of propellent gases are to be produced from 6 million tons of
coal. Based on present prices (April 1, 1981) coal hydrogenation
is not yet profitable either. From the estimated investment
costs of DM 4.4 billion, and the annual operation costs of roughly
DM 2.5 billion, a production price of 96 Pf/l can be calculated.
Including the costs for distribution, mineral oil tax and V.A.T.,
the selling price of petrol from coal would amount to DM 1.82
per liter.

In Table 3 the price structure of petrol produced from mineral
oil is examined and compared with that of coal. The price of raw
materials is comparable, although the heating price of crude
oil has in the meantime significantly surpassed that of coal.
This is because of the lower efficiency of hydrogenation: roughly
3 t of coal are necessary to produce 1 t of petrol. Especially
the processing costs, higher by a factor of 5, are very impres-

Table 3. Comparison of the petrol price from mineral oil and coal

	Gasoline from mineral oil (Pf/l)	Gasoline from coal (Pf/l)
Fuel costs	49.6	55.5
Operating costs	8.5	40.6
Distribution costs	14.0	14.0
Mineral oil tax	51.0	51.0
Vat	16.0	20.9
Petrol station price	139.1	182.0

sive. They characterize the fundamentally higher degree of difficulty connected with all coal processing techniques in comparison with the techniques for processing crude oil or even natural gas.

The costs for distribution and trade (line 3) are the same in both cases. The next two points are problematic and they need to be discussed in detail. In the first place, it is supposed that a mineral tax is also to be paid for coal derivates, and in the second place, it is supposed that, through V.A.T., the treasury makes additional profits from the higher degree of difficulties connected with coal processing techniques. The result is a selling-price ratio of DM 1.39 per liter compared to DM 1.82 per liter. The abolition of mineral oil tax alone would already lead to the same selling prices!

The price ratio between petrol from coal and from mineral oil has gone through an interesting development. In 1933, when it was still wise to do so, calculations based on world marketing prices resulted in a factor of roughly 2.8. In the 60s this rose to factor of 4. Further development since 1970 is shown in Fig.6. Without discussing the various influences individually, it can be stated that the devaluation of the dollar in 1977/78 caused a retardation. The basic tendency is, however, doubtlessly a falling one

These calculations are based on the assumption, as with coal gasification, that the production plants are equipped with aggregates for environmental protection according to environmental protection standards. These expenditures are very low considering the high investment costs for, contrary to the power plant, the amounts of flue gas emissions and sewage are very low and all processes take place in closed vessels.

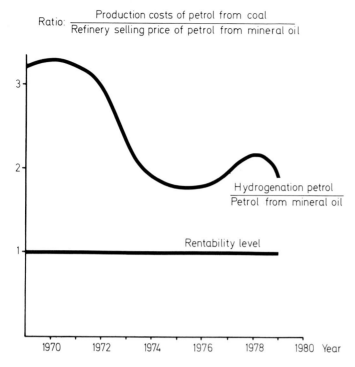

Ratio: $\dfrac{\text{Production costs of petrol from coal}}{\text{Refinery selling price of petrol from mineral oil}}$

$\dfrac{\text{Hydrogenation petrol}}{\text{Petrol from mineral oil}}$

Rentability level

1970   1972   1974   1976   1978   1980 Year

Fig. 6. Development of the production costs of petrol from coal in comparison to the re- finery selling price of petrol from mineral oil

4   Generation of Electricity and of Heat

The substitution of crude oil and natural gas by coal need not necessarily be done through hydrogenation or gasification. In some respects it seems more appropriate to do it through gener- ation of electricity and district heating. Therefore, these de- velopments will also be discussed here in brief.

The coal-fired power plant is economically competing with the nuclear power plant. Whereas in the basic load range light-water reactors are less costly than coal-fired power plants, the latter are superior in the medium- and peak-load range.

The utilization of coal-fired power plants for the medium- and peak-load range will, in the future, largely depend upon the technical perfection that can be achieved. Therefore, basically new techniques with a higher degree of efficiency, integrated en- vironmental protection measures, and lower power generation costs are being elaborated. In this connection, mention should be made of a promising new development, suitable for the gener- ation of electricity as well as heat in smaller units: the fluidized-bed process.

In the fluidized-bed process (Fig.7) fine-grained coal ash is placed above a distributor plate and combustion air is blown through it so that it remains suspended. Pneumatically added coal is burnt at a temperature of approximately 800°C, the generated heat being transferred to steam boiler pipes. The generated

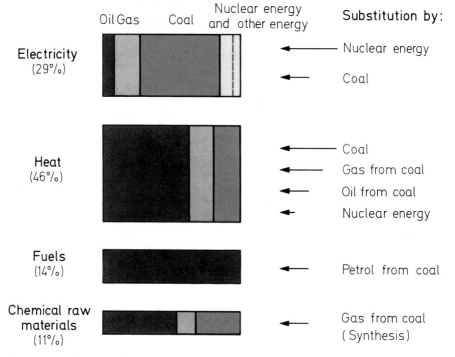

Fig. 10. Substitution of crude oil and natural gas by coal and nuclear energy

groups. The fields in the individual blocks show the momentary supply rate of crude oil, natural gas, coal, and nuclear energy.

The generation of electric power (approximately 30%) must be based mainly on nuclear energy. This is inevitable. Fuel oil and natural gas are to be eliminated and the lowest load range will have to be met by coal-fired power plants. Due to the exceedingly high rise in electricity consumption, also a considerably wider employment of coal is to be expected in the long-run.

The emphasis on crude oil substitution by coal will quite likely take place in the heating market, at present consuming nearly 50% of the primary energy used. Because of the higher degree of efficiency, it is advisable to employ directly coal here, e.g., by fluidized-bed coal combustion with combined power and generation heat, since the utilization of oil from coal and gas from coal would be energetically unprofitable by 50%. Only when the supply of natural gas decreases — at the turn of the century or even earlier — will it be possible to employ coal gasification with regard to a possible utilization of the already existing distribution network.

The other two fuel-consuming sectors, the fuel and the chemical sector, which together take up 25% of the energy market and are not as significant as often thought to be, ought to be supplied

with crude oil derivates as long as possible, even if the supplies of crude oil are diminishing. This makes a restructuring
of the refineries necessary. Only when quantities no longer suf-
fice can a large-scale hydrogenation of coal be considered. In
the chemical sector at least, the production of synthesis gas
produced from coal can be considered at an earlier stage, as
for example for methanol synthesis or the ammonia process. Synthesis gas from lignite is already profitable now. However, from
the political point of view, an earlier construction of a plant
with large hydrogenation capacity for fuels could be advisable.

6 Conclusion

The future availability of coal as primary energy carrier does
not only depend upon the resources at our disposal. Moreover,
possibilities of extracting them play a substantial role, too.

In the Federal Republic of Germany the geological reserves of
lignite and bituminous coal together amount to 250 billion tons
CE. Yet it can be seen that the fraction of national coal con-
tributing to the energy needs — lignite and bituminous coal —
will not exceed the momentary 30%, at most it might remain the
same. For one thing, no more capacity increases can be expected
from the lignite mining in the "Kölner Bucht," at present starting with the exploitation of the "Hambacher Forst" by means of
deeper pits. Despite the well-known and exemplary recultivation,
it seems, because of ecological reasons, impossible to raise the
amount of extracted lignite above 35 million tons CE.

The limits of exploitation have also been reached with bituminous coal. By digging new shafts and pits it may be possible
to raise the actual extraction capacity of 90 million tons by
some 10 million tons. Much more, however, will not be possible,
since it is problematic to continue aiming at rationalizing on
the one hand, and to seek additional laboure for coal mining on
the other hand.

When the processing capacities needed are available it will, in
the long run, be an inevitable necessity to import coal. For
this reason, the coal mining countries that come into question
should not only dispose of suitable mining capacities, but must
also have long-term access to the coal deposits. This can either
be in form of long-term supply agreements, or even better through
direct participation in the foreign mines. The world coal trade
of 280 million tons per years in 1980 will have to be considerably extended and needed transportation facilities will have to
be provided. There are indications that coal should only be imported in form of raw material, it could be directly hydrogenated
or gasified at the mines, so that the fluid products like oil or
methane can then be transported by tankers to the industrialized
countries. Coal processing techniques are more difficult and less
economical than the processing of crude oil and natural gas. We
will, however, be needing them, for natural coal resources are
practically inexhaustable. In addition, the coal deposits are of

easier access and less problematic from the political point of view, since they are more evenly spread over the whole world, which is not the case with crude oil reserves.

The Carbon Dioxide Problem

H. Oeschger, U. Siegenthaler, and T. Wenk[1]

1 Introduction

The climate of the earth depends to a great extent on the composition of the atmosphere. Atmospheric components play an important role regarding different kinds of energy transports (interaction with sunlight and thermal terrestrial radiation, atmospheric circulation) which determine the earth's heat budget and thus the climate.

As an example, carbon dioxide (CO_2) gas, though it constitutes at present only about 0.034% = 340 ppm (parts per million by volume) of the atmosphere, absorbs part of the thermal radiation emitted from the surface and keeps the surface temperature of the earth by several °C above the temperature we would observe for a CO_2-free atmosphere.

By the use of fossil fuels, as carbon, oil, and natural gas, huge amounts of CO_2 are being produced and emitted into the atmosphere. Estimates show that for an increasing consumption of fossil fuels, around the middle of next century the atmospheric CO_2 level might well reach the double of its original preindustrial value of 270 to 290 ppm.

Climate model calculations indicate that such a CO_2 doubling would lead to a rise of the average global temperature by 1.5° to 4°C, with an amplification by a factor of 2 to 3 in polar areas. Other gases released into the atmosphere due to human activities like chlorofluoromethanes, methane, and oxides of nitrogen have similar absorptive properties and might enhance the warming effect by up to 50%. Whereas the above mentioned gases interact mainly with the thermal radiation emitted from the surface, dust from volcanic eruptions reduces the amount of solar radiation being absorbed in the atmosphere and on the earth's surface, inducing a global cooling as observed after the Mt. Agung eruption in 1963. Anthropogenic aerosols released over cities and industrial areas, however, seem to lead to a local warming.

1 Physikalisches Institut, Universität Bern, Sidlerstr. 5, 3012 Bern, Switzerland

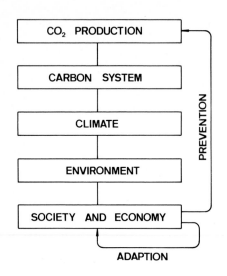

Fig. 1. CO_2 problem: The interconnection of different aspects

For a world with steadily increasing population, already now partly suffering from malnutrition, the perspective of a climatic change with probable impacts on food production is of considerable concern. For the assessment of the consequences of the anthropogenic CO_2 release qualitative and quantitative answers to the following questions are needed (Fig.1):

What will be the future trend of CO_2 production from fossil fuel consumption?

How will the released CO_2 be redistributed between atmosphere, ocean, and biosphere, i.e., how much will remain airborne?

How will an increased atmospheric CO_2 level affect global and regional climate?

What will be the effects of changing climate and environment on society?

At present, the answers one can give to the above questions are still afflicted with considerable uncertainties. Nevertheless, estimates on the nature and magnitude of CO_2-induced effects can be made, and compared with climatic changes in the past. This helps to foresee and minimize possible negative consequences.

In this article we attempt to review the present knowledge on different aspects of the problem and possible consequences for society.

2 The Global Carbon System

2.1 General Description

The chemical element carbon (C) plays an important role in a great number of processes:

In all biological processes, carbon compounds are formed, modified, or decomposed.

In the sea and in lakes, debris of organisms form sediment
layers which consist to a large part of carbonate ($CaCO_3$) and
organic carbon.

A tiny fraction of those sediments, reduced carbon compounds
in concentrated form, are our most important energy source:
coal, oil, and natural gas.

In the atmosphere, the CO_2 content significantly influences
the terrestrial radiation balance.

The behavior of carbon in the environment is very complicated.
Its general importance and man's interference, however, are mo-
tivation enough for a careful study.

Roughly, the exchanging carbon (Fig.2) can be divided into the
following four main reservoirs (all amounts are given by the
equivalent mass of pure carbon, so that the different carbon
compounds can be directly compared):

The *atmosphere* contains at present some 700 Gt (1 Gt = 1 Giga-
ton = 10^{12} kg) carbon in the form of CO_2. Its concentration has
been increasing since 1958 on the average by about 1.5 ppm per
year to a value of about 340 ppm in 1981.

In the *ocean*, carbon is present in different forms:

    as dissolved CO_2 (ca. 1%)
    as bicarbonate ions (HCO_3^-, ca. 85%)
    as carbonate ions ($CO_3^=$, ca. 10%)
    and also as dissolved organic carbon compounds (ca. 3%).

Altogether, the ocean contains roughly 40,000 Gt carbon, which
equals 60 times the atmospheric amount.

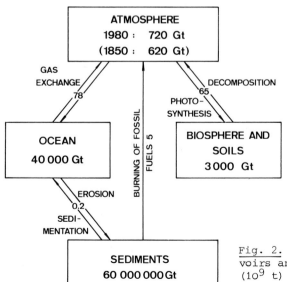

Fig. 2. Global carbon system main reservoirs and fluxes. Units are gigatons
(10^9 t) C for amounts, and gigatons C
per year for fluxes

Large regional differences make it difficult to estimate the carbon amount in the *land biosphere* (plants, humus and soils). Estimates range between 2000 and 3000 Gt, i.e., about three to four times the atmospheric content.

In *sediments*, carbon is stored in huge amounts, estimated at 70,000,000 Gt.

Of course, each of these main reservoirs represents itself a very complicated system and may be subdivided further.

Carbon is exchanged between the different reservoirs. The exchange fluxes try to bring the system to an equilibrium state, for which in each reservoir the influxes compensate the losses. For our purposes — the study of the fate of the CO_2 released due to human activity — it is very important to know the size of the different exchange fluxes. The biospheric fluxes, CO_2 assimilation and release of CO_2 by decomposition of organic matter, can be estimated based on ecological data.

The annual cycle of vegetation is reflected in the CO_2 record shown in Fig.3. During the growing season the removal of CO_2 from the atmosphere exceeds the production by decay of organic matter and the CO_2 level decreases by several ppm (depending on the geographical position of the site of observation), while in winter, when the biological productivity comes to a rest but decomposition continues, the CO_2 concentration increases again. Neglecting human interferences and possible natural changes in vegetation, on an annual mean, the total biomass remains constant, so that the biosphere represents neither a sink nor a source of CO_2.

In this context it is of interest to notice the coupling between the cycles of carbon and of oxygen. During photosynthesis, molecular oxygen is created, but an equal amount is used during de-

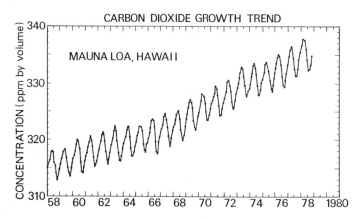

Fig. 3. Atmospheric CO_2 concentrations as observed at Mauna Loa observatory, Hawaii, by C.D. Keeling and co-workers (C.D. Keeling, pers. comm.)

composition of the produced organic matter, so that on average the biosphere is neither a source nor a sink for oxygen. There has to be expected, by the way, also no significant loss of atmospheric oxygen by the combustion of fossil fuels. Even if all available fossil energy resources were burnt, the amount of atmospheric oxygen would be reduced by not more than about 1.5% of its present value.

2.2 Determination of CO_2 Exchange Between Atmosphere and Ocean

^{14}C, a radioactive carbon isotope with a half-life of 5730 years, offers a possibility to determine the CO_2 exchange fluxes between atmosphere and ocean. It is produced in the atmosphere by the interaction of cosmic radiation with the atomic nuclei mainly of nitrogen and is chemically present as ^{14}C O_2. It follows the cycle of carbon and thus enters the vegetation as well as the oceans. When carbon is isolated from the atmosphere, its ^{14}C concentration starts to decrease due to radioactive decay. By measuring the remaining ^{14}C concentration it is possible to determine the time elapsed since isolation from the atmosphere.

There is a permanent CO_2 exchange between atmosphere and ocean. In the natural, unperturbed situation an equilibrium is assumed where the two fluxes, atmosphere to ocean and ocean to atmosphere, are equal. There is, however, a net flux of ^{14}C from air to sea, because ^{14}C decays in the ocean but is produced only in the atmosphere. This net influx must, in the stationary state, equal the radioactive decay in the ocean which makes it possible to calculate the exchange flux of CO_2, based on measurements of ^{14}C concentrations in atmosphere and oceans. The result is 78 Gt or about one-eighth of the preindustrial atmospheric CO_2 amount per year, in other words, the atmospheric residence time of carbon dioxide is 8 years.

If the atmospheric CO_2 concentration increases, the flux to the ocean increases proportionally. But as a reaction to the increasing carbon content of the ocean surface water, also the return flux grows; the resulting net flux has to be calculated by means of a model which takes into account the vertical ocean mixing.

From ^{14}C profiles it is known that the water in the deep sea below 1 to 2 km depth for the last time had been in contact with the atmosphere on the average about 1000 years ago. This means that water at that depth has not taken up fossil CO_2 up till now and will not do so in the next few decades.

2.3 The Airborne Fraction of Fossil Fuel CO_2

Since the beginning of industrialization approximately 180 Gt of carbon have been relased as CO_2 to the atmosphere. Not all this CO_2 remained in the atmosphere, a considerable part has been taken up by the ocean and possibly also by the biosphere. The fraction remaining in the atmosphere is called the *airborne fraction*.

Unfortunately, reliable measurements representative for the
global atmospheric CO_2 concentration have only been available
since 1958, when precision measurements at Mauna Loa (Hawaii)
and the South Pole were started by C.D. Keeling. It is therefore only for the past 20 years possible to calculate the airborne fraction. From the Mauna Loa record (Fig.3) we see that
from 1958 to 1978 the yearly average concentration increased
from 315 to 335 ppm. During the same time, an amount of CO_2 corresponding to 36 ppm has been released to the atmosphere due to
the combustion of fossil fuels. Thus, we get an airborne fraction F_a for this period of:

$$F_a = \frac{20 \text{ ppm (atmospheric increase)}}{36 \text{ ppm (fossil } CO_2 \text{ release)}} = 0.56$$

This means that 44% of the fossil CO_2 has been taken up by reservoirs other than the atmosphere. This estimate, however, is
afflicted with considerable uncertainties:

The CO_2 production of fossil fuels is not known exactly, the
uncertainty is of the order of ±5 to ±10%.

The deforestation of huge areas may be a considerable additional source for CO_2: wood and plant material is burnt or
decomposed. Some years ago, biologists estimated this CO_2
source to be about as great as the fossil CO_2 release (Woodwell et al. 1978). More recent estimates are not as high, but
they still indicate a significant CO_2 production from the
biosphere.

Little is known about the natural fluctuations of the atmospheric CO_2 level. There are indications for the existence of
such fluctuations as a result of anomalies in the atmosphere-
ocean exchange.

All this affects the estimation of the airborne fraction and
further investigations are needed to reduce the range of uncertainty. As the best estimate for the recent past we consider a
value of 56 ± 5%.

2.4 Models for Calculating Future Atmospheric CO_2 Concentration

If CO_2 production will continue to increase exponentially with a
growth rate of 4% per year, the airborne fraction will probably
remain in the range of 50% to 60% as observed in the last 20
years. For CO_2 production trends deviating from the exponential
growth, models are needed to calculate future atmospheric concentrations.

An important feature determining the airborne fraction is the
uptake of CO_2 by seawater at chemical equilibrium. Inorganic
carbon is present in seawater mainly as bicarbonate (HCO_3^-) and
carbonate ($CO_3^=$) ions, and to less than 1% only as dissolved CO_2.
An increase of CO_2 leads to a shift in the chemical equilibria
between these species, with the result that the excess CO_2 will
be partitioned between atmosphere and ocean in the ratio 1:6.
This means that even after a very long period, one-seventh of
the excess CO_2 will remain in the atmosphere. Chemical equilib-

rium with the whole ocean will be established, however, very slowly, so that the airborne fraction will be considerably larger over a long period of time. A number of carbon cycle models have been constructed which take into account the finite rates of air-sea gas exchange and of vertical ocean mixing. Their calibration is based on the ^{14}C distribution in atmosphere and ocean. Unfortunately the CO_2 uptake by the oceans cannot be measured directly because it is too small. However, a test is possibly by artificially produced ^{14}C and tritium which were released into the atmosphere by nuclear weapon tests, mainly in 1961/62. These isotopes have penetrated on average only a few 100 m into the ocean, so that we may conclude that also the anthropogenic CO_2, injected mainly in the past few decades, cannot have penetrated to much greater depth. The penetration of bomb-produced ^{14}C is relatively well simulated by existing carbon cycle models (Broecker et al. 1980).

In the following we present a typical model (Oeschger et al. 1975, Siegenthaler and Oeschger 1978) of the carbon cycle by discussing its responses to an assumed production stop and to a constant production rate.

In Fig.4 the computed CO_2 decrease after a supposed production stop in 1970 is plotted. (The result for a stop at present, 1981, or in future after a continuation of the exponential increase, would not be different in any essential way.) First the concentration decreases relatively fast and, after 30 years, the CO_2 excess is reduced to 70%. The decrease slows down and 50% of the original excess would be reached only around 2150.

If the production rate would be kept constant, say, again after the year 1970, one might first guess that the atmospheric CO_2 level would also reach a constant value for which production

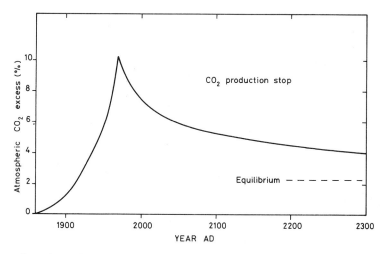

Fig. 4. Model-predicted atmospheric CO_2 concentrations for a hypothetical production stop in 1970

54

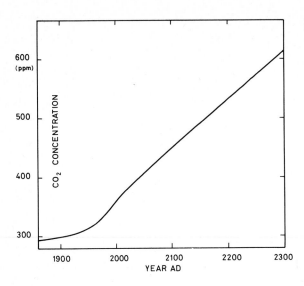

Fig. 5. Model-predicted atmospheric CO_2 concentration for
assumed constant production
rate after 1970

would be compensated by an equal flux into the ocean and perhaps
the biosphere. This is, however, not at all the case. Rather, the
atmospheric excess would continue to increase, after some time at
a nearly constant rate (Fig.5).

These two examples show the CO_2 release to be a real problem: In
contrast to, for instance, man-made dust which would disappear
within several years after its production had stopped, a considerable fraction of the cumulative CO_2 production will reside in
the atmosphere for very long periods of time.

3 The Climatic Impact on an Increased CO_2 Level

3.1 Modeling the Climate

When speaking about climate we have in mind the impressions of
weather manifestations, collected over a certain period of time.
How can climate be characterized in an objective way? Important
parameters are averages over several years of temperature, humidity, duration of sunshine, and precipitation rate, but also
daily and seasonal variations and frequencies of extreme situations such as droughts or floods. A great number of different
factors and processes influence climate on a large range of
scales in time and space, and it is impossible to simulate this
complex system even in a somewhat complete way. Climatologists
therefore try to take into account the most important mechanisms
in a variety of models of different complexity.

The simplest models describe the earth as an isothermal sphere
covered by an atmosphere (e.g., Rasool and Schneider 1971). More
complex ones imply a simulation of the latitudinal and vertical
dependence of temperature, of snow and ice cover, and of the heat
transport by atmospheric and oceanic circulation. The most sophisticated models, the *general circulation models* (GCMs), cal-

culate for a grid of points, distributed all over the planet, important climatic factors such as monthly averages of temperature, humidity, precipitations, or snow cover (e.g., Manabe and Stouffer 1980). They consider the distribution of continents and oceans and the most important physical properties of the earth's surface.

Surprisingly, the results of simple models agree quite well with those of complex general circulation models, but of course without regional resolution. Furthermore, calculated values of today's climate agree reasonably well with observation, supporting the validity of the model assumptions.

3.2 The Greenhouse Effect

The main factor determining global temperature is the solar radiation which heats the earth. Without reemission of energy, the temperature of the surface would increase continuously, but the surface emits thermal radiation at a rate proportional to the fourth power of its absolute temperature. Thus, the earth heats up to a temperature for which, on a time average, the absorbed solar radiation is compensated by emitted infrared radiation. Based on this equilibrium and taking into account that about 30% of the incident sunlight is not absorbed but reflected by clouds and the surface (Fig.6), an effective temperature (planetary temperature) of the earth can be calculated. This mean planetary temperature is 255 K or -18°C.

INCOMING SOLAR RADIATION	
REFLECTED IN THE ATMOSPHERE	26
REFLECTED AT THE SURFACE	4
ABSORBED IN THE ATMOSPHERE	20
ABSORBED AT THE SURFACE	50
ENERGY FLUXES FROM THE SURFACE	
IR REACHING SPACE DIRECTLY	10
IR ABSORBED IN THE ATMOSPHERE	105
CONVECTION	30
IR EMITTED IN THE ATMOSPHERE	
UPWARD	60
DOWNWARD	95

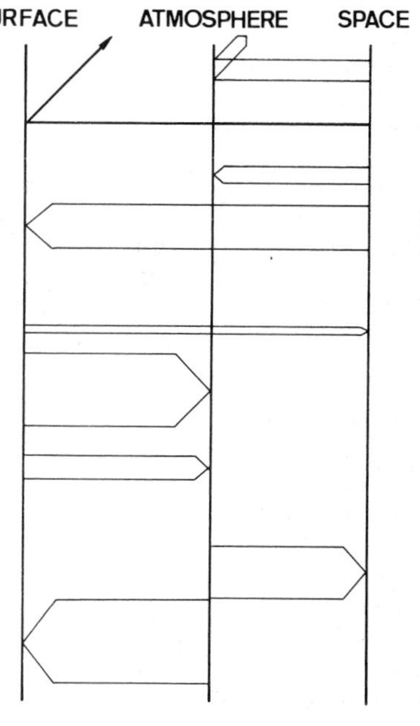

Fig. 6. The main energy fluxes between space, atmosphere, and earth surface. In equilibrium, influxes and losses are equal at each level. Incoming solar radiation at the top of the atmosphere: 100 units; IR, infrared radiation

The actual mean surface temperature is, however, about 15°C and thus considerably higher. The difference is due to the fact that most of the infrared radiation to space is not emitted by the surface but by higher (and colder) atmospheric layers, because — in contrast to the visible sunlight which penetrates the cloud-free atmosphere without much loss — the infrared radiation emitted by the surface is strongly absorbed, mainly by water vapor, but also by carbon dioxide, ozone and other trace gases. The absorbed energy heats the lower atmosphere which reemits infrared radiation, partly back to the surface. In this way the surface receives, in addition to the solar energy, a considerable amount of thermal radiation and is heated up to an average temperature higher than the effective planetary temperature. This phenomenon, that the atmosphere is transparent to the sun's radiation but absorbs in the infrared and thus acts as a spectrally selective thermal isolation, is called *greenhouse effect*.

The absorption of infrared radiation by water vapor leaves a spectral window at wavelengths between 9 and 15 μm. In this range, absorption occurs by CO_2 and other trace gases but not by water, and it is incomplete so that some radiation from the surface penetrates the atmosphere unaffected. An atmospheric CO_2 increase reduces the transmissivity in this spectral window and enhances the greenhouse effect. But also in other spectral regions, the greenhouse effect is amplified by a higher CO_2 concentration because the radiation emitted by the surface is absorbed more strongly and the back-radiation originates from lower, and therefore warmer, atmospheric layers. In this way, an increase of CO_2 (and other infrared-active trace gases) leads to higher surface temperatures.

Estimates show that for a doubled CO_2 concentration on the global average downward infrared flux increases by 4 Wm^{-2} (Augustsson and Ramanathan 1977). For the whole earth this corresponds to an additional energy flux at the surface of 2000×10^{12} W. This indirect heating due to the greenhouse effect can be compared with the direct energy input of 8×10^{12} W at present, showing that on a global scale the direct warming by energy use is much weaker than the CO_2-induced warming.

3.3 Results of Climate Models

To estimate temperature changes caused by increased atmospheric CO_2 concentrations one has to consider, beside the changed radiation balance due to CO_2 alone, a number of feedback mechanisms. A strong positive feedback is the increase in absolute humidity going along with the warming of the earth's surface and the lower atmospheric layers. The combined temperature effect of higher concentrations of CO_2 and of water vapor is two to three times that caused by CO_2 alone. Another positive feedback effect results from changes in the earth's reflectivity for solar radiation (albedo) owing to a decrease of the snow and ice cover in polar areas caused by the warming.

The effect of clouds is difficult to model. Increasing cloud cover on the one hand leads to a higher albedo but on the other

57

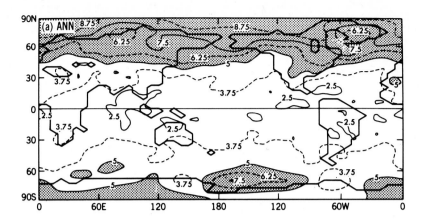

Fig. 7. Geographical distribution of the surface temperature increase caused by a quadrupling of CO_2, as calculated by a general circulation model. *Heavy lines* indicate contours of continents. An amplification in high latitudes and a north-south asymmetry are clearly visible. (From Manabe and Stouffer 1980)

hand also to an increase of thermal radiation. The two effects partly compensate each other. At present it is not known if a higher surface temperature will lead to a change of cloud cover.

Calculations by means of a variety of climate models of different complexity indicate for a CO_2 doubling an average global temperature rise by 1.5° to 4°C (WMO 1979), assuming that the energy fluxes at the earth's surface balance each other, i.e., that there is thermal equilibrium on a global scale.

While simple climate models provide estimates of CO_2-induced changes and permit to study the influence of specific processes and assumptions, reliable forecasts can finally only be expected from general circulation models. Figure 7 shows the rise of the mean surface temperature for a quadrupling of CO_2 according to a general circulation model (Manabe and Stouffer 1980). (The temperature differences for a CO_2 doubling are about half of those for a quadrupling.) The warming is about 3°C in equatorial regions and increases when going to higher latitudes; on a worldwide annual average it is 4.1°C. The amplification in polar regions is partly due to the snow-albedo feedback mentioned above. At high latitudes there is a seasonal asymmetry, with large temperature changes mainly in winter.

Temperature is not the only climate parameter of interest; changes in precipitation and evaporation are of equal significance and for some regions even more important. The results of Manabe and Stouffer indicate more precipitation in high latitudes. There are, however, still large problems with the simulation of the atmospheric water cycle which lead to uncertainties in the predictions.

3.4   Influence of the Oceanic Heat Capacity

All the climate model studies discussed so far assume that the
earth has adapted to the new radiative conditions and on average
has reached thermal equilibrium. In reality, however, the world
ocean has a considerable heat capacity and while the temperature
is rising, there is a relatively large flux of heat into the ocean needed to warm it up, so that the radiation budget is not balanced. Only very recently has this aspect, which causes a delay
in the global temperature increase, been included in the model
discussion in a quantitative way (Hansen et al. 1981, Hoffert et
al. 1980). Since heat is transported in the ocean by the same
processes as CO_2 — currents and turbulent water motions — the
same ocean models can be used for simulating the penetration of
the temperature increase into the ocean as for the CO_2 uptake.

Until 1980, the CO_2 increase was about 15%, and the corresponding model-predicted warming is 0.35° to 0.6°C (depending on the
model assumptions) if radiative equilibrium is assumed. Due to
the ocean's thermal inertia, the actual temperature increase is,
however, delayed by about 10 to 20 years, and the change until
1980 is, according to our model results, only 0.2° to 0.3°C.

Obviously, it would be desirable to have a check of these model
calculations by observed data. These data, however, indicate
that there are natural temperature fluctuations, the causes of
which are only partly known, that are larger than the presumed
CO_2 effect. Thus, the average temperature of the northern hemisphere increased by about 0.7°C from the middle of the last century until about 1940, then decreased again by about 0.3°C until
1960 or 1970. Probably, it will not be before the year 2000 that
a CO_2-induced warming will be distinguishable with some confi-
dence from natural variations.

The heat transport by ocean currents is undoubtedly important
for the distribution of climates on earth; the importance of the
Gulf stream for Western Europe is a good example. Unfortunately,
the present knowledge is insufficient to permit a realistic simulation of the three-dimensional ocean circulation and its influence on regional features of the anticipated climatic change.

3.5   Effects of Climatic Change

It is instructive to compare the possible future climatic changes
with climatic variations of the past (Flohn 1979). During the
last 100 years the temperature has remained within a band less
than 1°C wide, and during the little ice age (sixteenth to nine-
teenth century) it was about 1°C cooler in the northern hemisphere
than at present. Temperatures about 1° to 2°C higher than now
probably determined the altithermal period (ca. 4500-8000 years
ago). An increase of 4°C in global average temperature might produce an ice-free Arctic Ocean in summer, a situation that has not
occurred in the past few million years.

Changing temperature is not the only and perhaps not even the
most important effect which may occur. As a result of global

warming it is expected that climatic zones will shift poleward, which might for instance imply a modification of precipitation regimes in today's main agricultural regions, thus affecting global food production.

The strongly enhanced warming in polar areas may cause the Greenland ice sheet to shrink and some glaciologists emphasize that it could render the West Antarctic ice sheet unstable. These phenomena would lead to a rise of sea level by several meters affecting the population in coastal regions. The time scales for these processes are highly uncertain, but signigicant changes would probably not occur within the next 50 years.

In discussions of past climate a change toward higher temperatures is often considered as an improvement of climate. However, our society is strongly adapted to the present climatic conditions and any climatic change would force a readjustment to new environmental conditions in a relatively short time period, a process which could cause severe economic problems and might lead to social and political tension.

It has been mentioned that climatic changes caused by a CO_2 increase need not necessarily have negative effects only, but might bring along also beneficial consequences. If, however, developing regions, which already now are highly vulnerable, are affected negatively by climatic change, it is difficult to see how they could really benefit from positive effects in other parts of the world, unless a completely new global strategy of assessment and solution of problems would evolve.

4 Predictions of Future CO_2 Levels and Global Temperatures

Predicting CO_2 concentrations in the atmosphere requires the knowledge of future burning rates of fossil fuel. In view of the political and economic factors which influence fossil energy use, reliable predictions of the consumption are not possible, and it is not surprising that scenarios presented by different authors in the past few years vary within large limits.

We therefore only discuss an upper limit scenario, according to which all available fossil fuels will be burned, and a low-use scenario with a prescribed maximum CO_2 level.

Assumed CO_2 production rate (until 1980: given by energy use statistics) and model-predicted atmospheric CO_2 concentrations[2] for the upper limit scenario, which is based on total reserves corresponding to 8 times the preindustrial atmospheric CO_2 amount, are shown in Fig.8a. According to these assumptions, the CO_2 out-

---

[2] For calculating the CO_2 concentrations reproduced in Fig.8a, model parameters were chosen slightly different from those adopted for obtaining the curves of Figs.4, 5, and 9. Therefore, the results are not quantitatively comparable. The differences are, however, not very large and do not affect in any way the general conclusions

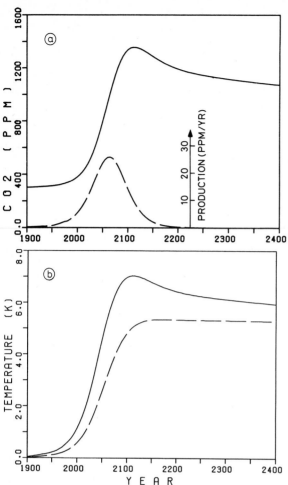

Fig. 8a,b. Upper limit scenario for CO_2 assuming that all fossil fuel reserves are burned. (a) Assumed production rate (*dashed line*) and the calculated CO_2 concentration (*solid line*). (b) Corresponding increase of global temperature, assuming radiative equilibrium (*solid line*), and considering the heat capacity of the ocean (*dashed line*)

put would reach a maximum, about 11 times the value of 1980, around the year 2060. The atmospheric CO_2 level would still continue to increase for about 50 more years to about 4.5 times the preindustrial value, then it would decrease. The decrease is, however, very slow, even after the virtually complete stop of CO_2 production, because it takes a very long time until the excess CO_2 reaches the deep ocean layers.

In Fig.8b the resulting temperature changes are plotted, based on the model of Augustsson and Ramanathan (1977) for constant cloud top temperature. The full line is valid for radiation equilibrium, while the dashed line was obtained taking into account the heat uptake by the oceans. The equilibrium curve indicates a maximum warming, around 2100, of 7°C, decreasing to 6°C until 2300-2400. If thermal inertia of the ocean is taken into account,

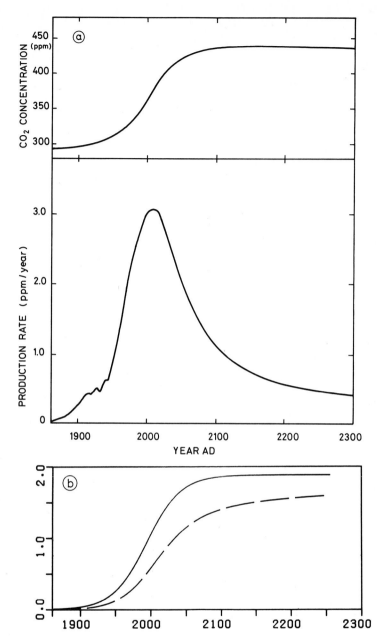

Fig. 9a,b. CO_2-limiting scenario. (a) Prescribed atmospheric CO_2 level, reaching at maximum 150% of the preindustrial value (*upper curve*); and production rate (*lower curve*), until 1970: based on fuel consumption, after 1970: adapted to the prescribed concentration trend. (b) Corresponding increase of global temperature, assuming radiative equilibrium (*solid line*), and considering the heat capacity of the ocean (*dashed line*)

the temperature increase would be less, a little above 5°C which is, however, still a rather large change. The reduction of the warming by the thermal inertia implies a certain mitigation of the climatic consequences of a CO_2 increase. On the other hand, the model results indicate that at least until 2400 A.D. the temperature would not decrease again, not even after the stop of CO_2 emissions. This finding, if correct, is a serious warning that when CO_2-induced climatic changes will be established as facts, the original situation will not be restored over a long period of time, not even by a complete stop of fossil fuel combustion. A CO_2-induced climatic change appears as essentially irreversible.

It may some time be reasoned that because of anticipated negative climatic consequences, the CO_2 level should not exceed a certain upper limit, for instance 150% of the preindustrial level.

The corresponding CO_2 production function (Fig.9a) computed by means of a CO_2 model, would allow for a moderate growth of fossil energy use until the end of this century, but then it would need to decrease relatively fast and drop below present values around 2030-2040 (Siegenthaler and Oeschger 1978). The final temperature increase for assumed radiative equilibrium with space would be between 1.2° and 1.9°C, depending on climate model assumptions (only one case is shown in Fig.9b) and would essentially be reache in about 2060. The thermal inertia of the ocean would cause an effective reduction of the increase by 15% to 25% in 2100, compared to radiative equilibrium, and even in 2250 still by 10% to 15%.

When considering the results discussed here, one should bear in mind that the models are affected with uncertainties. Therefore, the figures should not be interpreted literally, but rather as illustrations of the qualitative features of future CO_2 concentrations and temperatures.

5   Conclusions

5.1   International Activities

Since the middle of the last decade the CO_2 problem has received worldwide attention. It has been subject of a number of symposia and workshops.

National and international organizations are becoming active in the field of climatic change and its societal implications. Examples are the CO_2 program of the U.S. Department of Energy and the European Climate Programme of the Council of the European Community. At the beginning of 1980 the World Meteorological Organization (WMO) together with the International Council of Scientific Unions (ICSU) and the U.N. Environmental Programme (UNEP) implemented the World Climate Programme which includes subprograms devoted to obtain a better understanding of the physical basis for climate and climatic change and of the impact of changing climate on society.

5.2 Natural Sciences Aspects

The number of scientists working on aspects of the CO_2 problem has strongly increased and the state of knowledge has been critically reviewed.

Carbon cycle: Biologists pointed at the CO_2 input due to deforestation and oxidation of humus and corresponding uncertainties in the CO_2 balance. However, these uncertainties are not such that serious doubts of the knowledge about the main characteristics of the carbon cycle and its dynamics would be justified. It is improbable that the future CO_2 trend, at least for the next few decades, will deviate strongly from that calculated with present carbon cycle models.

CO_2-induced climatic change: The estimates of the equilibrium temperature increase for a CO_2 doubling, obtained by means of different models, essentially agree within a certain range. An aspect brought up recently and discussed in this article is the lag of the climatic response due to the ocean's thermal inertia. This effect leads to a moderate reduction of the maximum climatic excursion. Furthermore, the time at which climatic phenomena will be observed, which unambiguously can be attributed to enhanced CO_2 levels, will be delayed.

In spite of the complexity of the problem, the present scientific knowledge on its different aspects as summarized here is on solid grounds. If world fossil fuel use will continue to increase in the next 50 years, man will probably experience a significant global warming.

5.3 Socio-Economic Aspects

The socio-economic aspects of the CO_2 problem have been discussed intensively in the recent past (see, e.g., Kellogg and Schware 1981). Some of the ideas and conclusions which evolved from these discussion are:

For several reasons, as the delay of the climatic response, the virtual irreversibility of CO_2-induced climatic change and the time required to shift to non-fossil energy technologies, it seems unsuitable to wait with policy decisions and measures dealing with the CO_2 issue till climatic effects are observed.

In view of the difficulties in precisely predicting future CO_2 inputs and atmospheric concentrations and especially the corresponding regional climatic effects it seems advisable to provide policy makers and planners with scenarios of possible future climates in the different regions, including plausible changes in temperature, precipitation and evaporation, seasonal effects and general variability. Such scenarios could help to get a feeling for the possible impacts of the expected climatic change and the desirability of energy policy changes and measures to mitigate adverse impacts.

Averting a serious climatic change without greatly reducing fossil fuel consumption would imply removing CO_2 before it reaches the atmosphere or increasing the natural CO_2 sinks, e.g., by refor- estation. Ideas on CO_2 removal, e.g., by injection into deeper ocean layers, have been developed. They imply enormous costs for facilities and extra energy and are unattractive not only for economic reasons. Reforestation in itself seems useful but would require an increase of the gloval biomass by 1% each year to balance the present release of 5Gt per year, which appears to be unpracticable.

Thus the only effective measure is reduction of the fossil fuel consumption, at least to stop its growth. Whereas for the devel- oped countries energy conservation is feasible and can lead to some stagnation of fossil fuel use, this is not the case for the developing countries which need an energy growth for the improve- ment of their living conditions.

The replacement of fossil fuels by non-fossil ones, as nuclear and renewable energy sources, could help to keep the CO_2 increase in limits which correspond to an acceptable climatic change. As indicated in Fig.9a, a limitation of atmospheric CO_2 at 1.5 times the preindustrial value would be difficult to achieve and imply actions to restrict drastically the use of fossil fuels. There is a considerable probability that fossil fuel consumption in a few decades will lead to significant climatic changes, and at- tempts to decrease the vulnerability of the different regions and societies are highly desirable. Possible measures to improve resilience to climate change are protection of arable soils, im- proved water management, application of new agricultural tech- niques, and maintainance of global food reserves. From the point of view of improving the present living conditions, mainly for developing countries, such efforts are important already now.

The global aspects necessitate strong internationally coordinated scientific and technological efforts to create a world energy system which over a long period of time will not lead to environ- mental problems with concomitant undesirable socio-economic con- sequences.

References

Augustsson T, Ramanathan V (1977) A radiative convective study of the CO_2 climate problem. J Atmos Sci 34:448-451

Broecker WS, Peng T-H, Engh R (1980) Modeling the carbon system. Radiocarbon 22:565-598

Flohn H (1979) Can climate history repeat itself? Possible climatic warming and the case of paleoclimatic warmphases. In: Bach W et al. (eds) Man's im- pact on climate. Elsevier, Amsterdam Oxford London, pp 15-28

Hansen J, Johnson D, Lacis A, Lebedeff S, Lee P, Rind D, Russel G (1981) Climate impact of increasing atmospheric CO_2. Science 213:957-966

Hoffert MI, Callegari AJ, Hsieh C-T (1980) The role of deep sea heat storage in the secular response to climatic forcing. J Geophys Res 85:6667-6679

Keeling CD, Bacastow RB, Bainbridge AE, Ekdahl CA, Guenther PR, Waterman LS (1976) Atmospheric carbon dioxide variations at Mauna Loa Observatory, Hawaii. Tellus 28:538-551

Kellogg WW, Schware R (1981) Climate change and society, Westview Press, Boulder, Colorado

Manabe S, Stouffer RJ (1980) Sensitivity of a global climate model to an increase of CO_2 concentration in the atmosphere. J Geophys Res 85:5529-5554

Oeschger H, Siegenthaler U, Schotterer U, Gugelmann A (1975) A box diffusion model to study the carbon dioxide exchange in nature. Tellus 27:168-192

Rasool SI, Schneider SH (1971) Atmospheric carbon dioxide and aerosols: Effects of large increases on global climate. Science 173:138-141

Siegenthaler U, Oeschger H (1978) Predicting future atmosphere carbon dioxide levels. Science 199:388-395

WMO (1979) Report of the meeting of CAS working group on atmospheric carbon dioxide, Boulder, Colorado, Nov. 1979. World Meteorol Org, Geneva

Woodwell GM, Whittacker RH, Reiners WA, Likens GE, Delviche CC, Botkin DB (1978) The biota and the world carbon budget. Science 199:141-146

74

Table 2. Technical data of typical power reactors [9,10]

		PWR Biblis B	BWR Krümmel	HWR Atucha 2	Magnox Wylfa	AGR Heysham 1	HTR Uentrop	FBR Super Phenix
Capacity, gross electr.	MW	1300	1316	745	655	666	308	1240
" net "	MW	1240	1260	692	590	622	296	1174
" thermal (core)	MW	3733	3690	2160	1876	1510	750	3000
(net) efficiency	%	33,2	34,1	32	31,4	41,2	39	40
Type of fuel		UO_2	UO_2	UO_2	U	UO_2	UO_2/ThO_2 (coat.part.)	UO_2/PuO_2
Heavy metal content (U, Th, Pu)	t	102,7	155,8	85,1	595	110	7,4	32
Enrichment	%	3,0	2,6	nat.	nat.	2,1-2,6	93	14,5/18,5
Average fuel rating rel. to heavy metal	MW/t	36,4	23,7	25,4	3,15	13,7	101	93,8
Average fuel rating rel. to fissile material	MW/t	1212	911	3525	444	536	1157	600
Core, height × diameter	m×m	3,90×3,64	3,71×4,99	5,30×6,06	9,10×17,4	8,2×9,3	6×5,6	1×3,66
Average power density	MW/m³	92	50,9	14,1	0,87	2,71	5,08	275
Average burn-up rel. to heavy metal	MWd/t	32500	27500	7500	4600	18000	100 000	65 000
Conversion factor		≈0,55	≈0,6	≈0,8	≈0,8	≈0,6	≈0,7	≈1,2
Shape of fuel assembly		16×16-20	8×8-1	Cluster 37	Individ. rods	Cluster 36	Ball	Cluster 91
Fuel rod diameter	mm	10,75	12,5	12,9	27,9	15,3	(coat.part.)	8,5
Cladding tube material/wall thickness	mm	Zry4/0,72	Zry2/0,85	Zry4/0,57	Magnox	steel/0,38	Graphite	steel/0,7
Max. fuel temperature	C	2430	1900	2250	569	1500	1250	
Coolant; pressure	bar	H_2O; 155	H_2O; 70	D_2O; 115	CO_2; 28	CO_2; 41	He; 40	Na; 2,5
Inlet/outlet temp.	C	290/323	215/286	278/312	247/414	287/651	250/750	395/545
Flow rate	t/h	72 000	55 600	37 080	36 948	13 272	1 062	59 760
Moderator		H_2O	H_2O	D_2O	Graphite	Graphite	Graphite	–
Control rods: material/number		AgInCd/61	B_4C/205	HfSt/18	BSt/185	BSt/81	B_4C/78	B_4C
Steam conditions before turbine: pressure/temperature	bar/C	52/266	67/282	55,9/271	45,9/401	160/538	180/530	180/490

Fig. 8. Biblis nuclear power plant with two pressurized water reactors

1 reactor pressure vessel
2 steam generator
3 reactor coolant pumps
4 internal concrete cylinder
5 annulus ventilation system
6 containment
7 secondary containment
8 safety injection pump
9 residual heat removal pump
10 residual heat exchanger
11 borated water storage tank
12 accumulator
13 intermediate cooler
14 service cooling water pumps
15 main steam and feed water
    valve compartment
16 emergency generator
17 demineralized water
    storage tank
18 emergency diesel engine
19 fuel pool pump
20 nuclear component
    cooling pumps
21 emergency nuclear component
    cooling pumps
22 emergency service cooling
    water pump
23 extra borating pump
24 rapid shutdown system

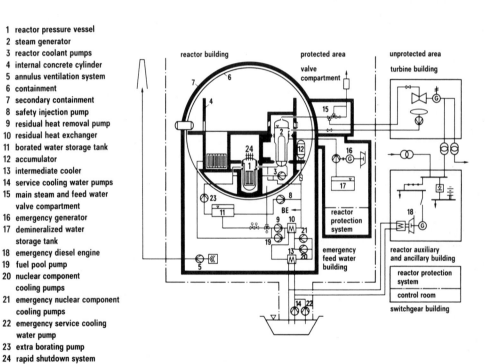

Fig. 9. Engineered safety features (ESF) of KWU pressurized water reactors

All pressurized primary system components are located in a cylindrical or spherical reactor building (Figs.8 and 9) which serves
as safety containment (in the Eastern bloc countries, this is only
the case with the new 1000 MWe plants).

A nuclear explosion is impossible in principle with a reactor, as the chain reaction would automatically stop as a result of negative temperature and density coefficients of reactivity. In the case of a loss of coolant, too, the PWR will automatically shut down due to a lack of moderation (so-called inherent safety). The radioactivity remaining in the fuel assemblies, however, represents a considerable potential danger. After a prolonged operating period, the activity inventory of a large PWR is about 10^{10} Ci [15,16], and the so-called residual heat output resulting from it is a certain percentage of the original reactor output. Therefore there is the danger of fuel-rod overheating and release of radioactivity during a loss of coolant.

The main task of the extensive engineered safety features (ESF) is therefore to ensure reactor cooling during all conceivable off-normal events. The engineered safety features of a large pressurized water reactor (Fig.9) comprise for example:

> eight accumulators (12) which contain borated water and, in the event of a large loss of coolant, feed quickly and automatically sufficient water into the coolant loops via check valves (8 × 20% redundancy);

> four low-pressure residual heat removal pumps (9) which, during a loss-of-coolant accident, pump borated water from the borated water storage tank (4 × 50%), or from the building sump if the tanks have already been emptied, into the primary system via residual heat exchangers, and which ensure the cooling of the reactor core even after the accumulators have been emptied in the event of a large leak (redundancy of 4 × 50%);

> four high-pressure safety injection pumps (redundancy of 4 × 50%); emergency power supply consisting of batteries and motor-generator sets (redundancy of 2 × 100%) and quick-starting diesel units (redundancy of 4 × 50%) and other equipment.

The containment is designed to absorb primary coolant mass and energy even in the event of a total coolant loss (so-called design basis accident).

The safety design and assessment of a plant are based on an exact analysis of all possible accident sequences and impacts, i.e., of all conceivable chains of events under consideration of the relevant probabilities of occurrence. It is not possible to discuss the various comprehensive studies on the risks of a power plant in detail here (see for example [16,17]); however, they have shown that the risks caused by nuclear power plants are very small compared to other risks associated with everyday life. Especially the probability of serious accidents happening is extremely low. These results are confirmed by experience: Up to 1981 i.e., during a total of 1200 reactor operating years, nobody outside a nuclear power plant has been injured by a nuclear power plant. (This applies also for the frequently quoted accident in the TMI-2 nuclear power plant in Harrisburg.) According to all previous knowledge and experience, pressurized water reactors are not only reliable and economic but also safe and environmentally acceptable energy supply plants.

2.2.2 Development Trends

Except for nuclear ship propulsion units, pressurized water reactors have mainly been used for electricity generation in large power plants to date. This will continue to be the main field of application in the future. The main concern of the manufactures at present is a consolidation of the status achieved until now, i.e., a standardization of the power plants and their components to simplify the licensing and evaluation procedures, to shorten the construction periods, to increase reliability, safety and availability, and to improve their economics. The present standard plants have a capacity of 900 to 1300 MWe, a certain flexibility being possible especially by component standardization. Thus, for example, plants with three or four largely identical reactor coolant loops can be offered. A development of even larger unit capacities up to 2000 MWe would not pose any problems with respect to technology, but it is not of topical interest for the time being. However, safety and environmental aspects have led to a still on-going discussion on new types of construction, e.g., underground construction. With ground embedment of a power plant, the protection against plane crashes and blast waves, for example, would be greater than with the usual type of construction; but, on the other hand, safety-related drawbacks with respect to operation, inspection and possibly repair of certain components are to be expected [18,25].

In addition to electricity supply, the supply of heat for industrial purposes and for district heating systems may become important in the future.

Nuclear heating and power plants with pressurized water reactors are technically feasible without any major problems. Figure 10

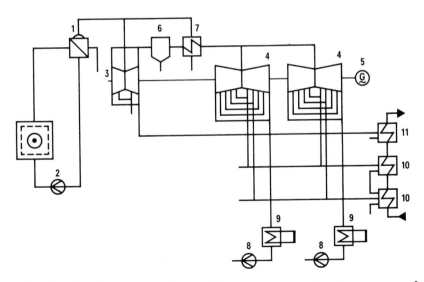

Fig. 10. Heating power plant with a pressurized water reactor [18]. 1, Steam generator; 2, reactor coolant pump; 3, HP turbine; 4, LP turbine; 5, generator; 6, moisture separator; 7, thermal superheater; 8, condensate pump; 9, condenser; 10, hot-water preheater; 11, peak-load heater

shows the single-line diagram of a nuclear district heating and power plant with a pressurized water reactor; its design includes a multi-stage hot-water heating system. Heating steam is tapped from various expansion sections of a condensing turbine. If a high heating output is required, the peak load heater may be cut in as a topping unit. If standard condensing turbines are used, this arrangement permits to achieve heating capacities of up to about 1200 MWth by minor changes on the turbine casing with 1300 MWe plants. The reduction of the electrical capacity is relatively low. For example, with a heating capacity of 1000 MWth and a heating water temperature of 165°C, it is about 160 MWe [18]. Compared to the costs of the distribution systems, the energy costs are almost negligible [19]. For example, the capacity of the nuclear power plant in Grafenrheinfeld would be sufficient to supply the whole city of Schweinfurt (56,000 inhabitants) about 10 km away, with district heat. However, the economic question still remains whether the installation of the piping system is worthwhile, especially in the case of the mostly large distances between the supply point and the users.

This applies similarly to the consumption of industrial heat. Here, too, most of the required amount is used for space heating while the remaining share is process heat, most of which can in turn be supplied in the form of process steam by pressurized water reactors [18,19]. For brown coal gasification, for example, up to 80% of the heat is required for heating and drying the coal and for steam generation at temperatures below 300°C so that it is possible to use the PWR as heat source if only the heat of reaction proper is generated by burning coal. Compared to autothermic coal gasification, high-temperature reactors would save about 40% of the required coal, and pressurized water reactors about 25% [18].

In addition to power plants with heat extraction, special heating plants which will not generate any electricity at all might become interesting. The French "Thermos" project may be cited as an example. This is a pressurized water reactor with an integrated primary system (heat exchangers and pumps are inside the reactor pressure vessel), which is designed for a thermal capacity of 100 MW, a service water temperature of 120°C and a primary-side coolant pressure of 8.5 bar [23].

Another aim of development is the improvement of fuel utilization. While usual pressurized water reactors have a conversion factor of about 0.55 (ratio of generated to spent fissionable nuclei) and utilize only about 0.6% of the natural uranium if there is no reprocessing and plutonium recycling, the utilization of natural uranium could be raised to about 0.9%, i.e., an increase of about 50% by plutonium recycling already technically proven in service. A further improvement of uranium utilization (by a factor of 3) could be achieved if the volume ratio between moderator and fuel (V_{H_2O}/V_{UO_2}) of about 2 (for the PWR) is reduced to a value around 0.5 and if PuO_2/UO_2 fuel with a content of about 7.5% of fissionable plutonium is used. Thus the conversion factor is increased to about 0.95. The result is a high conversion rate APWR (APWR = advanced pressurized water reactor).

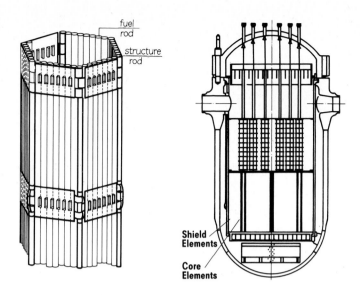

Fig. 11. Fuel assembly (*left*) and reactor pressure vessel (*right*) of an Advanced Pressurized Water Reactor (APWR)

Moderation ratios of such low values, however, cannot be achieved with the known square fuel-rod lattices (see Fig.7), but only with hexagonal rod arrangements as they are known in principle from the fast breeder reactors. Fuel reprocessing is required, too. Due to the highly reduced rod-to-rod spacings, such a high conversion rate APWR of about the same specific fuel-rod power has a considerably lower core height (Fig.11). All external components such as pumps, steam generators and others, however, could remain substantially unchanged [20,21,30].

2.3 Boiling Water Reactors

The boiling water reactors which are also cooled and moderated by means of water (H_2O) differ from the pressurized water reactors above all by the fact that the water in the reactor boils. In all more recent BWRs, the saturated steam generated in the reactor is passed directly into the turbine, by which means a simplification of the cooling and working cycle is accomplished (Fig.12): The large steam generators and the pressurizer are dispensed with; the recirculation capacities are lower. The reactor operating pressure (about 70 bar) is lower than that of the PWR (see Table 2), with the operating temperature being about the same.

A disadvantage of the BWR is the working cycle activation which, although controllable, is not quite avoidable.

The reactor core design is similar to that of the PWR. Here, too, slightly enriched UO_2 in Zircaloy cladding tubes is used as fuel, but the achievable heat flux densities on the fuel-rod surface are somewhat lower due to the required DNB margin[1] and the specific

---

1 DNB = Departure from nucleate boiling

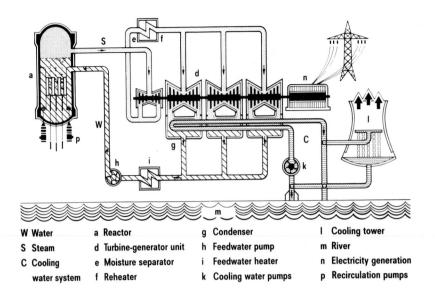

W Water	a Reactor	g Condenser	l Cooling tower
S Steam	d Turbine-generator unit	h Feedwater pump	m River
C Cooling	e Moisture separator	i Feedwater heater	n Electricity generation
water system	f Reheater	k Cooling water pumps	p Recirculation pumps

Fig. 12. Conceptual flow diagram of a nuclear power plant with a boiling water reactor (source: KWU)

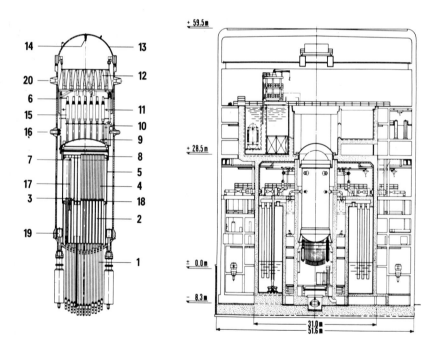

Fig. 13. Reactor pressure vessel (*left*) and reactor building (*right*) of a boiling water reactor (KWU). *1*, Control rod drives; *2*, control rod guide tubes; *3*, lower edge of active zone; *4*, fuel assemblies; *5*, core shroud; *6*, pressure vessel; *7*, upper edge of active zone; *8*, top guide; *9*, shroud head; *10*, core spray line; *11*, normal water level; *12*, steam dryer; *13*, pressure vessel closure head; *14*, head spray cooling system; *15*, cyclones; *16*, feedwater sparger ring; *17*, control rods; *18*, core plate; *19*, axial-flow reactor internal pump (RIP); *20*, steam outlet nozzle

fuel ratings (23.7 compared to 36.4 MW per t uranium according
to Table 2) are smaller. Fuel inventory and core volume are higher
than with the PWR for the same capacity.

Despite a lower operating pressure the reactor pressure vessel
is considerably longer and heavier than that of the PWR because
the space abouve the core is required for steam separators and
because the control rods must be inserted from the bottom (Fig.
13). Besides, the number of the control rods is higher as (different from the PWR) liquid boric acid cannot be used for reactivity
control due to the two-phase flow.

In most of the cases, the reactor building (see Fig.13) of boiling water reactors is equipped with a so-called pressure suppression system. During a loss-of-coolant accident, the escaping steam
is not accommodated by the containment (as it is the case in the
PWR), but it is passed through vent pipes into a water-cooled
chamber and condensed there. The pressure suppression system permits a smaller reactor building to be designed, but the thermo-
hydraulic processes are more complicated than in the full-pres-
sure containment.

In terms of cost, the advantages and disadvantages of the boiling water reactors seem to counterbalance each other. In the past,
the ratio of PWRs to BWRs ordered remained nearly constant even
over a prolonged period of time (see Fig.3). During the last few
years, however, an increasing lead of the PWR has emerged which
ultimately is due to the clearer separation of radioactive and
non-radioactive power plant areas.

The BWR power plants which are under construction or being planned
at present have capacities of 900 to 1300 MWe. Lower capacities,
as they are required in the developing countries for example, are
considered to be not economical. Recently, however, a new BWR
design was presented by KWU which shall permit the economic generation of power and process heat even with capacities of 200 to
400 MWe [22]. It is a dual-cycle BWR with natural circulation of
the coolant in the reactor pressure vessel and natural recirculation of the condensate from the steam-to-steam heat exchanger. The
steam-to-steam heat exchanger separates the radioactive primary
system from the non-active secondary system (as is the case in the
PWR), pumps being largely avoided. The relatively simple design
justifies the expectation of low capital costs and high availability and safety.

2.4 Heavy-Water Reactors

Heavy water = D_2O ($D = {}_1^2H$ = deuterium), 0.015% of which is contained
in natural water, differs from H_2O by a considerably lower neutron
absorption and a lower moderation. Therefore, D_2O, which can be
produced by isotope separation, is a favorable, though very expensive, moderator. Its low neutron absorption permits the utilization of natural uranium (0.72% U 235) instead of enriched uranium
(about 3% U 235 for PWR) required for H_2O moderation. On the other
hand, a considerably higher amount of D_2O than H_2O is required
due to the lower moderation capacity: The required moderator/fuel

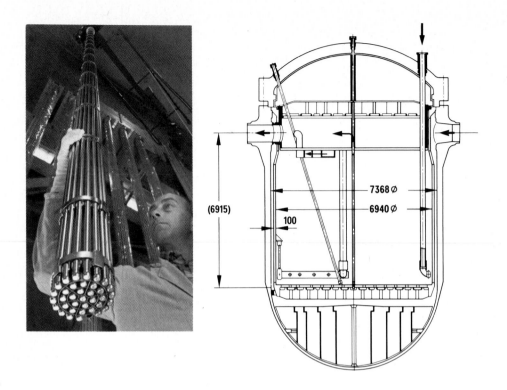

Fig. 14. Fuel assembly (*left*) and reactor pressure vessel (*right*) of a D_2O-moderated pressure vessel reactor (KWU)

volume ratio is about 20 to 1 (instead of 2 to 1). Therefore the volume of D_2O reactors compared to that of H_2O-moderated reactors is larger by the factor 10 for the same capacity. Capital costs are always higher.

The fuel-rod design is similar to that for light-water reactors: i.e., UO_2 pellets in Zircaloy cladding tubes. As, however, far less D_2O is required for cooling purposes than for moderation purposes, the fuel rods are not distributed equally over the entire reactor (as is the case in light-water reactors with their contiguous fuel assemblies), but arranged as fuel-rod clusters in separate cooling tubes which are surrounded by D_2O (Fig.14). A typical rod cluster consists, for example, of 37 rods having a diameter of about 1 cm, and has a cluster diameter of 10 cm, and a centerline spacing of 30 cm (see data listed in Table 2).

The functional separation of moderator (D_2O) and coolant permits the use of other coolants (e.g., CO_2 or H_2O steam) but most of the D_2O-moderated reactors are also cooled by D_2O.

There are two different designs of the D_2O reactors, i.e., the pressure vessel and the pressure tube reactors.

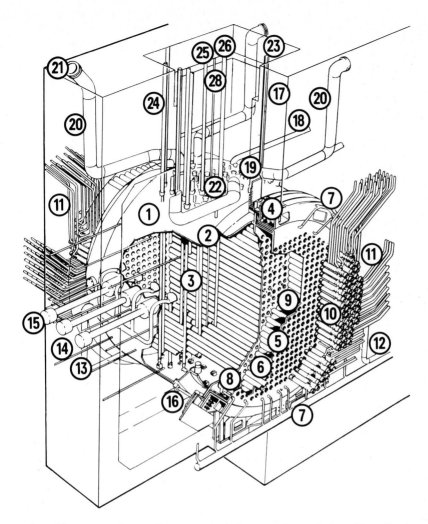

Fig. 15. D$_2$O-moderated pressure tube reactor of Canadian design (Candu type) [33]. 1, Moderator tank (calandria); 3, pressure tubes; 6, pressure tube extensions; 12, moderator outlet; 13, moderator inlet; 24, 25, 26, control rods

In the pressure vessel reactor (KWU's design series), cooling tubes and moderator region are contained in a common pressure vessel. Coolant and moderator have the same pressure. The pressure vessel dimensions, however, are considerably larger than for a conventional PWR. For example, the pressure vessel of the 745 MWe Atucha II nuclear power plant (see Fig. 14 and Table 2) has an outer diameter of about 8 m and a mass of about 1000 t [32]. Up to a capacity of 1000 MWe, the fabrication technology of D$_2$O reactors does not pose any problems. Refuelling is quasi-continuous during power operation. The refueling machine is located above the reactor pressure vessel. Therefore the control rods are inserted into the reactor core from top in an inclined position (see Fig. 14).

In the pressure tube reactors (of the type designed by the Canadian AECL), the cooling tubes are designed as pressure-retaining tubes penetrating the practically pressureless moderator tank (calandria) over its entire length. The Canadian pressure tube reactors form a horizontal cylinder with horizontal cooling channels. The cooling channels are accessible for refueling machines at both front faces. The control rods are inserted from the top (Fig.15).

The advantage of the pressure tube design is that size and capacity of the reactor are not limited by the pressure vessel dimensions; on the other hand, however, there is more neutron-absorbing material in the reactor core which is exposed to high mechanical stresses.

Table 2 contains important technical data of the 745 MWe Atucha II heavy-water reactor. Compared to PWR power plant, the following differences can be seen:

> lower thermal efficiency

> related to the heavy metal inventory (uranium inventory), the fuel rating is lower, however, related to the fissile material inventory (U 235 inventory), it is considerably higher. The difference is due to the use of natural uranium

> power density in MW per m^3 of core volume is considerably lower

> the attainable burnup (MWd/t of uranium) is considerably lower, as the initial excess reactivity of natural uranium is low

> considerably higher conversion factor.

To date, heavy-water reactors have not been able to establish themselves in industrialized countries with the exception of Canada, as their higher capital costs are not compensated by correspondingly lower fuel costs. However, D_2O reactors are still an interesting alternative, as natural uranium can be used as fuel, and as enrichment plants are not required.

2.5 Graphite-moderated Reactors

In terms of neutron physics, graphite differs from D_2O by a slightly higher neutron absorption and a slightly lower slowing-down power. Therefore, graphite-moderated reactors require even higher moderator/fuel volume ratios than D_2O reactors and are therefore even larger (the capacity being the same). Natural uranium may also be used as fuel, but then the attainable burnup is even lower than with D_2O reactors so that enriched fuel is used for the more recent types.

Until the mid-60s, graphite was the prevailing moderator substance (see Fig.3). And the first nuclear reactor in the world, designed by E. Fermi in Chicago in 1942, was a graphite-moderated "uranium pile." Today, in substance, four main types are differentiated: Magnox reactors, advanced gas-cooled reactors (AGR), high-temperature reactors (HTR), and graphite-moderated light-water-cooled reactors (LWGR).

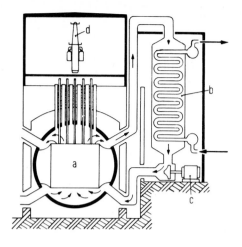

Fig. 16. Fuel element (*left*) and conceptual arrangement (*right*) of a Magnox reactor [12,15]. *a*, Core; *b*, steam generator; *c*, circulator; *d*, refueling machine

2.5.1 *Magnox Reactors*

This type of reactor (Calder Hall type) prevailing until the mid-60s was developed in Great Britain and, in a modified design, also in France. Individual fuel rods of natural, metallic uranium in Magnox cladding tubes (= magnesium alloy), having a diameter of about 3 cm, are used as fuel.

The fuel rods are arranged in holes spaced at about 20 cm of the graphite moderator assembled in blocks (Fig.16). CO_2 is used as coolant. Technical data are contained in Table 2.

Due to the relatively low maximum cladding tube temperature of about 450°C (Magnox!), lower heat flux densities and larger moderator volume, the fuel ratings are very low; the same applies to the average power density in the core (nearly 1% in comparison with that of the PWR). As a result, capital costs are high. Despite a higher steam temperature, the net power plant efficiency is lower than that of the PWR because of the high auxiliary power required by the CO_2 circulators. As the burnup (MWd/t) is lower than that of the PWR by a factor of 7, the low fuel fabrication costs cannot compensate the other disadvantages. Magnox reactors cannot compete with modern PWRs and BWRs and are no longer constructed. However, at present they still constitute a substantial part of the British nuclear power plant capacity and may be considered as the forerunners of the AGR and the HTR.

2.5.2 *Advanced Gas-cooled Reactors (AGR)*

Advanced gas-cooled reactors are a further development of the Magnox type by the British. Moderator and coolant (CO_2) were retained; however, to increase outlet temperature (efficiency), fuel rating (power density) and burnup, rod cluster elements with enriched UO_2 (2% to 3% U 235) in cladding tubes of stainless

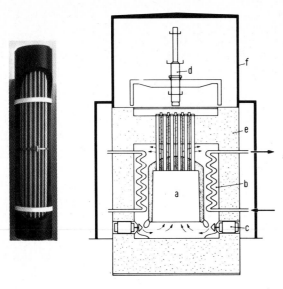

Fig. 17. Fuel element (*left*) and conceptual arrangement (*right*) of an Advanced Gas-Cooled Reactor (AGR) [12,15]. *a*, Core; *b*, steam generator; *c*, circulator; *d*, refueling machine; *e*, prestressed concrete vessel; *f*, containment

steel (Fig.17) are used. Steel is resistant to higher temperatures but it is less favorable than Magnox in neutron physics terms. Such fuel elements permit CO_2 outlet temperatures of about 650°C.

The reactor is located in a cylindrical prestressed concrete pressure vessel (see Fig.17), to be found in a similar form already with the more recent Magnox reactors. Steam generators and CO_2 circulator are integrated in the vessel wall. As superheated steam is generated, a conventional turbine generator unit can be used. This is doubtlessly a great advantage. Low power density, high heat capacity, and small maximum leak cross sections are also favorable characteristics in terms of safety engineering. A comparison of the most important technical data reveals considerable improvements over the Magnox reactor. Compared to the PWR, however, the following disadvantages can be seen:

  lower fuel rating by a factor of 3, i.e., the correspondingly larger fuel inventory;

  lower power density by a factor of 30 (capital costs!)

  distinctly lower burnup

  poorer accessibility of important components in the prestressed concrete vessel.

To date, reactors of this type have been constructed only in Great Britain. But there, too, their economics and competitiveness with pressurized water reactors are controversial.

2.5.3 High-temperature Reactors (HTR)

This type of reactor is characterized by coolant outlet temperatures above 700°C. Such temperatures cannot be achieved with CO_2 cooling and metal cladding tubes. Therefore, HTRs have a helium cooling system and completely ceramic fuel elements. At present, this type of reactor does not play any part in the energy economy (see Fig.3 and Table 2). In the longer term, however, high-temperature reactors are promising especially in the process heat utilization field. They have also favorable characteristics in terms of safety engineering. Details may be seen in the next chapter of this book.

2.5.4 *Light-Water Graphite Reactors (LWGR)*

Graphite-moderated and light-water-cooled pressure tube reactors were developed exclusively in the U.S.S.R. and, together with the pressurized water reactors, are the basis of the Soviet nuclear energy program today.

The newer LWGR reactors have electrical and thermal capacities of 1000 MW and 3200 MW, respectively. UO_2 with 1.8% enrichment and contained in Zircaloy cladding tubes is used as fuel. Fuel inventory is about 190 t, the achievable burnup is 18,500 MWd/t. Steam with a pressure of 65 bar and a temperature of 280°C is generated [10,24]. The initially proof-tested nuclear steam superheating was abandoned at least provisionally. This type of reactor is not constructed in the Western world.

Besides the reactor types dealt with here, the only type of fundamental importance is the fast breeder (see Table 1). It will be discussed later in this book; technical data, however, are contained in Table 2.

3 Conclusions and Future Prospects

The preceding descriptions should have shown that there are great differences between the various types of reactors. Apart from power generation by nuclear fission, a pressurized water reactor and a Magnox reactor or a HTR, for example, have hardly anything in common.

From the physical point of view, reactors or nuclear power plants having a low neutron loss, those with good fuel utilization and a high efficiency should be the most advantageous ones. In practice, however, low capital costs, low fuel costs, and high reliability and availability are more important aspects. Physical, technical, and economic criteria are by no means identical. It is therefore not surprising that at the beginning of nuclear technology development about 25 years ago, most different types of reactors were conceived. Only during the last 10 to 15 years, the light-water reactors (PWR and BWR), the physical characteristics of which are by no means optimal, have gained a distinct lead (see Fig.3), but they have not been able to oust the other types completely. The decisive factors for this success should ultimately be the relatively uncomplicated design and, as a result, low capital costs. Therefore the lead of pressurized water reactors will probably even increase in the next 10 to 20 years. But what is the future of nuclear energy itself?

The future development of nuclear engineering will depend above all on the worldwide energy demand, the available energy resources, and the acceptance of nuclear energy by the public.

88

The problems of energy demand and energy resources are dealt with
in detail elsewhere in this book. There exists general agreement
on the fact that fossil energy resources will be exhausted in the
foreseeable future. Crude oil and natural gas deposits will last
for a few decades, coal deposits at most for a few centuries. Re-
generative energy sources such as sun and wind presumably will
be able to play only a complementary role, at least in central
Europe, in the last analysis because the available power densities
are too low.

But the "range" of the uranium deposits is also limited. Taking
into account the uranium reserves presently known, of the order
of 5×10^6 t, the current nuclear power generation of 10^8 MWthd/a
(corresponding to 140,000 MWe), and the natural uranium requirement typical of light-water reactors of 1.8×10^{-4} t/MWthd, our
uranium reserves would still last for about 280 years. Assuming
a nuclear power output of about 2000 GWe to be expected in the
year 2000, the "range" of the reserves is reduced to 20 years.

If the total nuclear energy was generated by heavy-water reactors
(without fuel reprocessing), natural uranium utilization and its
"range" would be improved by 30% to 50%. A hypothetical combination of conventional PWRs and high conversion rate APWRs (at the
ratio 1 to 2) would lead to an increase of the "range" by a factor of 3. A decisive improvement of uranium utilization by a factor of about 60 can only be achieved by fast breeder reactors.
The corresponding "range" would then be at least thousands of
years as uranium ores having lower uranium contents could be utilized commercially. In the longer term, fast breeders are indispensable due to their breeding properties. However, the capital
costs of fast breeders are considerably higher than those for
light-water reactors so that their "market prospects" depend mainly on the date of a beginning uranium shortage or price increase.

Fast breeder reactors as well as high conversion rate pressurized
water reactors, if any, require fuel reprocessing, as their advantages take effect only with the recycling of the bred plutonium
Fuel reprocessing, however, is in the last analysis not only required for reasons of energy economy but is also advantageous for
reasons of ecology, as the radiotoxicity of nuclear waste is reduced by the removal, reprocessing, and use of the plutonium.

While the technical and commercial potential of nuclear energy is
hardly contested, there is now as ever strong opposition against
a further development of nuclear technology in some countries.
The most important reasons are probably the rejection of large-
scale technology for ideological reasons and the fear of injuries
to health. These concerns refer in substance to radioactive emissions from nuclear plants, the nuclear waste disposal problem,
and the possibility of serious accidents.

There is no reason for the fear of radioactive emissions during
normal operation of nuclear power plants. As already explained
in Sect.1, these emissions are so insignificant (<1 mrem/a) compared to the natural background radiation exposure (average value
about 110 mrem/a) and its local variations (about ±50 mrem/a)
that there is hardly another type of energy generation which is

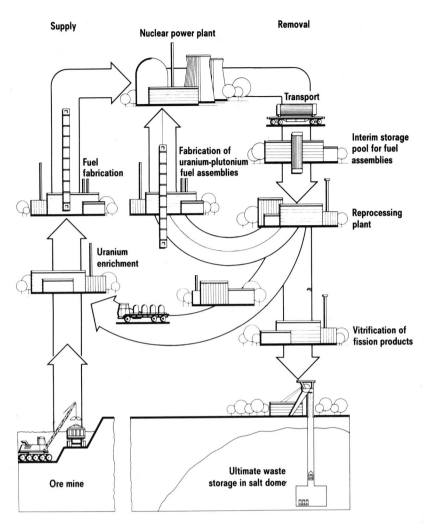

Fig. 18. Nuclear fuel cycle (source: DWK)

more environmentally compatible. I cannot see any unsolvable problems in nuclear waste disposal either. Reprocessing of nuclear fuel is possible on an industrial scale and implemented at present with success especially in France (see for example [26]); the same applies to the vitrification of high-level waste. Although the ultimate storage of this waste, for example in salt domes or granite formations, has not yet been proved on a large scale, no prohibitive problems are to be expected judging from the investigations and calculations carried out so far (see chapter of M. Rapin in this book). The amount of high-level waste is low: about 3 m^3 vitreous mass are produced from the waste of one 1200 MWE PWR per year, i.e., about 1 mg vitreous mass per kWh. Furthermore, an ultimate storage facility is a passive structure without any moving parts in which only slow solution and diffusion processes are possible at a low pressure and a comparably low temperature so that even in the event of unforeseen events, there is a lot of time left for the initiation of countermeasures.

In this respect, there is no doubt that the conditions are less favorable in nuclear power plants. Although the probability of serious accidents with large-scale environmental effects is very low in accordance with previous knowledge and experience, a certain "residual risk" of catastrophes cannot be precluded. The public mind set on this residual risk is not justified by actual facts (minor incidents, which are more probable, are more important as a whole) but it is psychologically understandable. In this context, some latest investigations [27,28], the results of which are considerably more favorable with respect to the possible maximum extent of an accident, are therefore of considerable interest. One of the facts explained in [28] is that in the risk analyses available the spread of important radioactive isotopes after a possible failure of the containment structure was overestimated and that the effects of even the most serious accidents are relatively low due to physical processes: "Whether an accident does or does not occur depends on our skill, although some like to think of it in terms of luck or probability. But the consequences of such an accident is not a question of skill, or luck, or probability — natural processes will limit the dispersal of significant radioactivity to the near vicinity of the accident. As a result, a public catastrophe will not occur." If this opinion should prove to be right, the actual (quite acceptable) total risk will hardly be changed, as it is not caused by spectacular large-scale events but rather by the much more frequent minor incidents; however, the most important psychological argument against nuclear energy, i.e., the "residual risk" would lose much of its importance. Therefore, one may look forward to further discussions with keen interest.

Acknowledgments. The author would like to thank the Kraftwerk Union AG for providing figures and in particular Mr. W. Henschel, Dipl.-Ing., for translating the manuscript into English.

References

1 Jahrb Atomwirtsch (1981) Handelsblatt GmbH, Düsseldorf
2 Interatom (1981) Energie für morgen
3 Oldekop W (ed) (1979) Druckwasserreaktoren für Kernkraftwerke, Thiemig, München
4 Münch E (ed) (1980) Tatsachen über Kernenergie. Verlag W Girardet, Essen
5 Kolb W, Die Emission radioaktiver Stoffe mit der Abluft aus Kern- und Steinkohlekraftwerken. Physikalisch-Technische Bundesanstalt, Ber PTB-Ra-8
6 Der Bundesminister des Innern, Umweltradioaktivität und Strahlenbelastung, Jahresbericht 1977, Bonn 1980
7 Kempken M (1980) Verzeichnis der Kernkraftwerke der Welt. Atomwirtschaft-Atomtechnik 25:467
8 Strahlenschutzverordnung der Bundesrepublik Deutschland (1976) Anlage IV
9 Oldekop W (1977) Leistungsreaktoren, Allgemeines. Ullmans Enzyklopädie der technischen Chemie, Vol 14. Chemie, Weinheim
10 Power Reactors (1980) Nucl Eng Internatl Vol 25, No 302
11 Stauber E (1977) Sicherheitstechnik und Umweltschutz bei Kernkraftwerken, Sonderdruck der KWU

12 Gruhl H, Kison H, Nieder R, Goetzmann C, Preuss H-J (1977) Thermische Reaktoren. Ullmans Enzyklopädie der technischen Chemie, Vol 14. Chemie, Weinheim

13 Holzer R, Konstruktion des Reaktorkerns, see [3]

14 Putschögl G, Die Reaktorstation in Biblis, see [3]

15 Oldekop W (1975) Einführung in der Kernreaktor- und Kernkraftwerkstechnik, Thiemig, München

16 Der Bundesminister für Forschung und Technologie, Deutsche Risikostudie (1979) TÜV Rheinland

17 Smidt D (1979) Reaktor-Sicherheitstechnik. Springer, Berlin Heidelberg New York

18 Preuss H-J, Entwicklungstendenzen und Zukunftsaussichten (des DWR), see [3]

19 Keller W (1975) Energieversorgung durch Leichtwasserreaktoren, Atomwirtsch Atomtech Okt 1975:476

20 Hennies HH, Märkl H (1980) Überlegungen zur Modifizierbarkeit eines LWR im Hinblick auf eine bessere Uranausnutzung. Jahrestagung Kerntechnik, Berlin

21 Berger HD, Oldekop W (1981) Neutronenphysikalische Aspekte eines fortgeschrittenen Druckwasserreaktors (FDWR). Jahrestagung Kerntechnik, Düsseldorf

22 Voigt O (1980) 200 MW-Kleinkraftwerk mit Naturumlauf. Atomwirtsch Atomtech 25:615

23 Labrousse M et al. (1978) Nucl Techn 38

24 Power Reactors in Member States (1979) IAEA, Wien

25 Keller W, Belda W (1981) Stand der Entwicklung von Kernkraftwerken mit Leichtwasserreaktoren, Atomwirtsch Atomtech 26:157

26 Scheuten GH (1981) Die Entsorgung der deutschen Kernkraftwerke, technisch gesichert, politisch ungesichert. Jahrb Atomwirtsch

27 Hennies HH, Hosemann JP, Mayinger F (1981) Ablauf und Konsequenzen eines DWR-Kernschmelzunfalls. Atomwirtsch Atomtechn 26:99

28 Levenson M, Rahn F (1980) Realistic Estimates of the Consequences of Nuclear Accidents, EPRI 9403

29 Atw-Schnellstatistik, Kernkraftwerke (1980) Atomwirtsch Atomtech 26:211

30 Zeggel W, Neelen N, Nissen K (1981) Thermohydraulische und konstruktive Randbedingungen eines fortgeschrittenen Druckwasserreaktors (FDWR). Jahrestagung Kerntechnik, Düsseldorf

31 Seifritz W (1980) Sanfte Energietechnologie-Hoffnung oder Utopie? Vol 92. Thiemig, München

32 Nuclear power plant with pressurized heavy water reactor. Kraftwerk Union, April 1980

33 Cahill L et al. (1974) Gentilly 2. Nucl Eng Internatl 19:481

34 Levenson M, Rahn F (1981) Realistic Estimates of the Consequences of Nuclear Accidents, Nucl. Techn. Vol. 53 May 1981:99

High-temperature Reactors[1]

R. Schulten[2]

It is a generally realized fact that the structure of energy technology is changing, mainly characterized by the fact that crude
oil, the principal raw material for generating energy is being
substituted by other long-term energy sources. In Fig.1 the his-
toric development of energy consumption in fractions of primary
sources of energy is depicted. One can see that, starting with
the consumption of renewable energy sources in the late middle
ages, first coal, then crude oil and gas were and still are the
most important energy resources of mankind, and that, starting
from the year 2000, a new energy technology will obviously be
needed to be able to meet future energy requirements. These connections are, however, not only characterized by the fact that
new forms of energy have to be found but also that the required
amounts of energy are by far more than those consumed till now,
meaning that a completely new problem arises here. How much more
energy will exactly be required can be easily and clearly assessed
there being no need of complicated and intricate system analytic
statements to be able to predict these quantitative changes. It
is generally accepted that the worldwide population will considerably increase (up to 8-12 billion) in the future and that the
living standards of the future population must at least be raised
somewhat in comparison with the present average living standard
in order to guarantee a stable worldwide social order for the
future. Even under the optimistic assumption that the worldwide
population will only reach 10 billion and that every person would
on the average only consume 50% of the energy now being consumed
by a citizen of the Federal Republic of Germany, approximately
30 billion t/CE per year will be needed to meet future requirements. Compared to the present energy production, this is tant-
amount to an increased energy production by approximately a factor of 4. This extraordinarily difficult problem does not belong
into the distant future but rather is already discernible, so that
appropriate solutions will have to be worked out during the next
20 years at the latest. The problem is, however, not just one
concerning the large amounts of energy; there is also the prob-

---

1 This paper was translated by Mrs. S. Messele-Wieser

2 Institut für Reaktorentwicklung der Kernforschungsanlage Jülich GmbH,
   Postfach 1913, 5170 Jülich, FRG

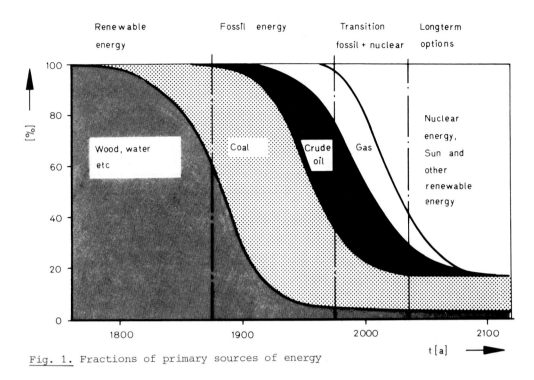

Fig. 1. Fractions of primary sources of energy

lem of bringing, to a tolerable extent, the high rate of energy
consumption in accordance with environmental conditions. In addition, basic energy sources such as crude oil, gas, coal,
uranium, and thorium are not evenly distributed over the whole
world. Because of geological circumstances, dependencies and
difficulties with the distribution have resulted and will remain
in the future as long as man is not capable of creating a uni-
versal government that could take charge of a fair compensation
and balancing of energy consumption. Considering the figures
stated, the estimated global amount of crude oil reserves now
appears to be gigantic, but it is not the solution for future
energy problems. There is an estimated amount of 300-400 billion
tons of crude oil. The price of this is three times higher than
at the moment and it is distributed in the world in such a way
that in the future dependencies will ensue. Referred to the required output, the amount stated will, theoretically, only suf-
fice to cover the global energy requirements for a couple of decades. The simple figures must inevitably lead to the conclusion
that new energy sources will be needed after the year 2000, so
that due to the long time needed to introduce new types of
energy this problem already has to be tackled now.

A further important point concerns the structure of energy requirements, in other words, in which form will these large quantities of energy presumably be available. Here we differentiate
between the market for electric power and compare it with the
market for energy for fuels and raw materials. In most industri-

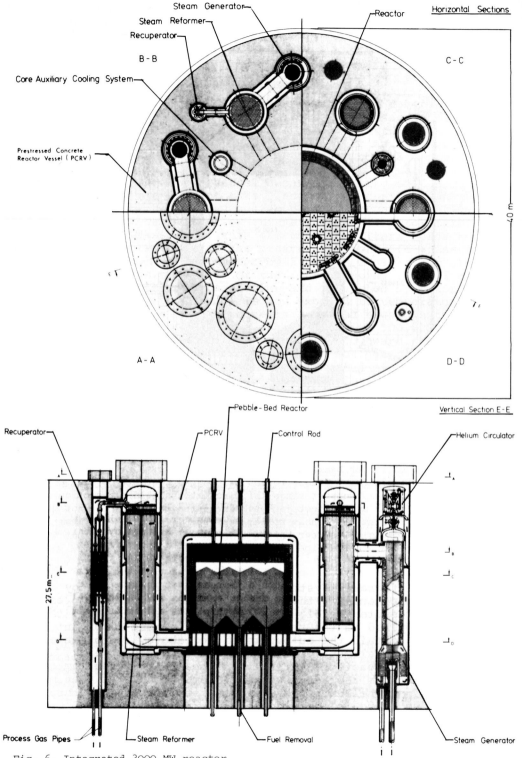

Fig. 6. Integrated 3000 MW reactor

lator returns the cooled helium to the components mentioned above
and to the reactor, so that all metallic parts of the components,
except for the heat exchange tubes of the steam reformer and the
steam generator, work at a temperature of less than 300°C. The
shutdown system, the circulator, and the system for loading and
removal can be taken over from the THTR. The most important new
component is the steam reformer that is used for the process of
hydrogenating coal gasification. The design and construction of
these new components is similar to the corresponding conventional
plants used in the chemical industry. The main difference compared to the conventional systems is the fact that helium is
being used as a heat transfer medium at a pressure of 40 atm,
whereas conventional plants operate with flue gas at a pressure
of 1 atm. This mainly means fewer material problems, since no
temperatures above 950°C can occur. During the last 4 years parts
of this steam reformer have already been tested in technical experiments. For instance, a number of pipes from the EVA plant
(see Fig.7) were tested at a temperature of 1000°C and a pressure
of 40 bar. A number of material samples for the steam reformer
are being tested in long-term experiments of extensive material
testing programs.

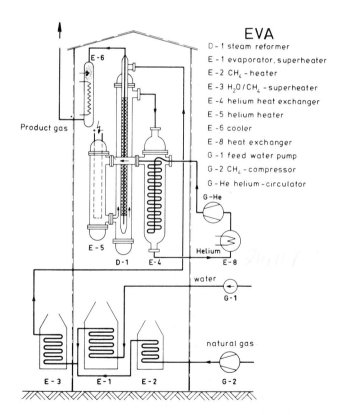

Fig. 7. Simplified flow sheet of the pilot plant EVA single-tube experimental plant

102

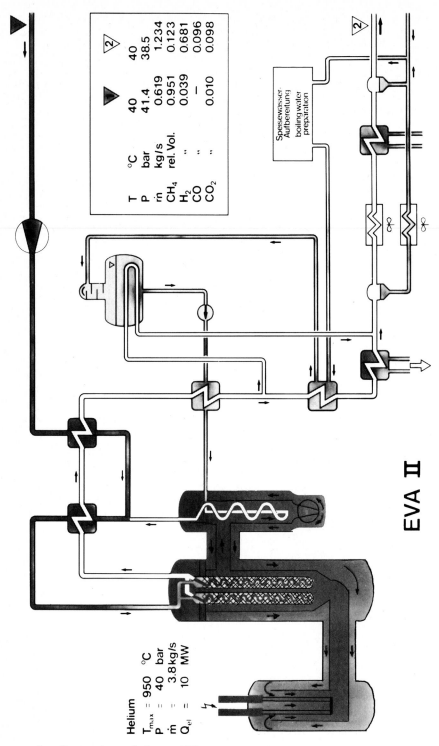

			▶	2
T	°C		40	40
P	bar		41.4	38.5
ṁ	kg/s		0.619	1.234
CH_4	rel. Vol.		0.951	0.123
H_2	"		0.039	0.681
CO	"		–	0.096
CO_2	"		0.010	0.098

Speisewasser-Aufbereitung
boiling water preparation

EVA II

Helium
T_{max} = 950 °C
P = 40 bar
ṁ = 3.8 kg/s
Q_{el} = 10 MW

Fig. 8. Design of Super-EVA

During the last year the construction of the Super-EVA has been started (see Fig.8) to demonstrate a characteristic bundle of the steam reformer system in long-term operations.

4   Process Application

In the project "Prototyp Nukleare Prozeßwärme" (prototype nuclear process heat) of the Federal Republic of Germany, two coupling cycles of nuclear plants with subordinate processes have been developed. One process consists in the reforming of methane with the help of steam and nuclear heat, the other consists in the gasification of coal with the help of steam and nuclear heat. Methane reforming can be used for a number of processes. The simplest process is the hydrogasification of coal. Since 3 years a semi-technical plant is in operation to demonstrate this process. In an exothermic reaction hydrogen and coal are transformed into methane through a well-known process. Half of the produced methane is then transported to the nuclear system's steam reformer where it is transformed into carbon monoxide and hydrogen and then completely converted into hydrogen. The hydrogen thus produced can then be reconducted to the coal hydrogasifier. The reaction schematic is shown in the following scheme:

$$4H_2 + 2C \longrightarrow 2CH_4$$
$$CH_4 + H_2O \longrightarrow CO + 3H_2$$
$$CO + H_2O \longrightarrow CO_2 + H_2$$

The same method can be applied for the liquefaction of coal. In the Bergius-Pier process, developed already 40 years ago, hydrogen and coal is, at a high pressure, brought to react in an oil sump. Besides the hydrocarbons a number of light, gaseous hydrocarbons is produced. These are transported to the steam reformer where they are reformed with nuclear heat and steam. The gases produced through this reforming process are converted into hydrogen. The reaction schematic of this is as follows:

$$<C> + <H_2> \longrightarrow C_nH_{2n+2} \text{ (gasoline, oil)}$$
$$+ C_iH_{2i+2} \text{ (hydrocarbons, gas)}$$
$$C_iH_{2i+2} + iH_2O \longrightarrow iCO + (2i + 1)H_2$$
$$iCO + iH_2O \longrightarrow iCO_2 + iH_2$$

total: $<C> + <H_2O>$ (gasoline, oil) + CO_2

In a similar process methanol can also be produced from oil. In this case 50% of the gas from the steam reformer is used for the hydrogasification of coal, the other 50% is used to produce methanol. Methane, the product of hydrogasification of coal, is returned to the steam reformer for the production of cracked gas, so that new cracked gas is either available for the production of methanol or for the hydrogasification of coal. The schematic of this reaction is here shown:

$$6C + 12H_2 \longrightarrow 6CH_4$$
$$6CH_4 + 6H_2O \longrightarrow 6CO + 18H_2$$
$$2CO + 2H_2O \longrightarrow 2CO_2 + 2H_2$$
$$4CO + 8H_2 \longrightarrow 4CH_3OH$$

Total $6C + 8H_2O \qquad 4CH_3OH + 2CO_2$

The steam reformer can, however, also be applied for transferring large amounts of energy in gas pipes. For this the chemical equilibrium of methane reforming is taken advantage of. Through the nuclear heat resulting from the reforming of methane, roughly 50 kcal/mol enter the gas cycle. After transmission of the cracked gases these can, at a lower temperature, be transformed again whereby the enclosed energy is again released. In this way it is possible to supply a number of consumers with heat energy, of a temperature of up to 500°C, from one large plant. This system is especially suited for the generating of process heat and household heating.

The direct gasification of coal by using steam is shown in the following scheme:

$$C + H_2O \longrightarrow CO + H_2$$

The temperature needed for the reaction depends upon the type of coal being used and amounts to 750°C-900°C. Altogether, approximately 30 kcal/mol are needed. The resulting gas mixture of carbon monoxide and hydrogen, usually also containing methane, has first to be processed. The processed gases can be utilized to produce natural gas and fluid hydrocarbons. The advantage of this process compared to conventional gasification processes lies in the fact that the amount of coal needed per generated heat unit is lower than a factor of 1.6. The balance of produced carbon dioxide, too, is lower by factor 1.6. Extensive studies concerning the economic of these processes have shown that, based on the coal prices in the Federal Republic of Germany, they can, in the long run, compete with crude oil and natural gas provided that realistic prices are fixed for crude oil, natural gas, and coal. One important result being that large plants with a capacity of approximately 3-4 million t/HCU/a show no need for essentially higher investment costs than those of conventional autothermic gasification plants as far as their capacity and the produced amounts of natural gas or fluid hydrocarbon are concerned.

In this year we have for the first time succeeded in experimentally realizing a promising procedure that allows the utilization of nuclear heat to decompose water into hydrogen and oxygen. This is a two-phase process; the first step is an electrolysis with a molten salt electrolyt at a temperature of 400°C. Here hydrogen is chemically bound in a chemical reaction so that the necessary amount of electricity can be reduced by a factor of 3 compared to the present electrolytic process. The removed hydrogen has then to be released out of the chemical compound by means of high-temperature treatment. Experimental results available till now indicate that such a chain of reactions can allow an efficiency of more than 55%. This is twice as much as the degree of efficiency being achieved through electrolysis at the moment. We presume that hydrogen thus produced will be able to compete with the expected future prices of fossil energy carriers. The produced hydrogen and oxygen will, apart from being used directly, possibly also be of importance in the coal technology and for the processing of heavy oil fractions. The production of hydrogen from nuclear heat is very valuable since a number of our present problems could thus be solved. After the complete development of this reactor and its fuel cycle, the primary initial fuel will mainly consist of thorium, so that, in the long run, independence of uranium imports can also be reached. The easy transportability of hydrogen and oxygen by means of gas ducts can afford more flexibility concerning the choice of site. In the last analysis hydrogen is the only kind of fuel that will allow solving all future environmental problems, since when it is combusted only water is produced.

There are also plans concerning the application of nuclear heat for producing oil from oil shale. Preliminary experiments of this kind have already been made. Some countries, especially developing countries with large oil shale resources, are interested in such a development.

At present only a few of these techniques are profitable in comparison with the prices of oil and gas. Studies indicate that the gasification of lignite using nuclear heat would already be interesting at the present if nuclear heat were already available. The studies furthermore indicate that a relatively negligible price increase for crude oil of 1%-2% above the overall inflation rate could have the effect that all techniques mentioned above would at once become economically profitable.

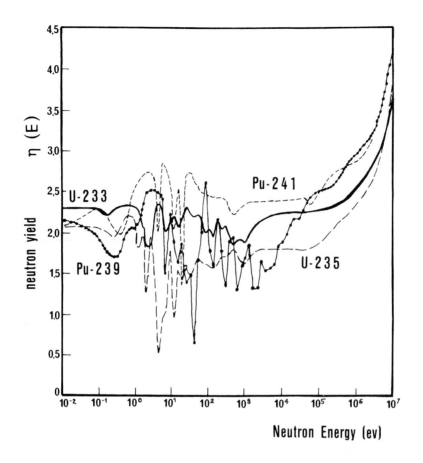

Fig. 2. Neutron yield for different fissile heavy isotopes as a function of neutron kinetic energy

etc.) by neutrons with energies >1.2 MeV (fast fission effect). The neutron yield η is the dominating quantity in the relation for the conversion and breeding ratio, respectively. Its components for individual isotopes are represented in Fig.2 as a function of the kinetic energy of neutrons for the fissile U-233, U-235, Pu-239, and Pu-241 fuel isotopes. It is seen that Pu-239 and Pu-241 assume the highest possible η-values in the region >5 × 10^4 eV (50 keV). U-233 has higher values than all the other fissile isotopes (η = 2.28) mainly in the range of thermal energies (10^{-2} eV), whereas U-235, the only non-artificial fissile material, mostly attains lower values than the other fissile fuel isotopes.

Typical values of (a + ℓ) are 0.3-0.45 in the reactor core and the surrounding breeding blanket of an FBR. The fast fission factor f can reach approximately 0.1-0.15. In this way, working in the uranium/plutonium fuel cycle with Pu-239 and Pu-241 as the fissile materials and U-238 and Pu-240 as the fertile materials, one attains the breeding ratios referred to above of

BR = 1.15-1.30. In the thorium/uranium-233 cycle, the attainable
breeding ratios would be lower. This explains the general preference for the U/Pu cycle in FBR's, with the start-up plutonium
(first core) of breeder reactors initially coming from converter
reactors. At breeding ratios of BR = 1.15-1.30, FBR's produce
some 100 kg of Pu/GWe·a above and beyond their own consumption.
This excess plutonium can either be used to build up first cores
of new FBR's or it can be burned in converter reactors. If no
excess plutonium is desired, the breeding ratio can be reduced
to a level around 1.

During power production nuclear fission generates fission products, and structural materials sustain material damage by fast
neutrons. Therefore, the fuel elements of FBR's must be reprocessed chemically and refabricated after having been in the reactor core for approximately 3 years. The fission products and
higher actinides, such as americium and curium, are chemically
separated and new steel structures are inserted. In chemical reprocessing and refabrication, approximately 1% of the plutonium
and uranium go into the waste as losses sustained in the process.
This means that after repeated passage through the fuel cycle
only 60%-70% of the fuel can finally be utilized in FBR's. While,
above a certain minimum breeding ratio, this fuel utilization in
FBR's is a function only of the burnup of the fuel in the reactor core and of the losses in the fuel cycle, it is mainly the
conversion ratio which is of decisive importance in converter
reactors (Fig.3). Converter reactors with relatively high para-
sitic neutron absorption (LWR's) only have fuel utilizations of
0.6%. Certain converter reactors (such as HWR's and HTGR's),
optimized in the core neutron physics to relatively high conversion ratios around 0.9, can attain fuel utilizations of a few
percent. FBR's, however, achieve fuel utilizations of 60%-70%
with breeding ratios around 1.15-1.30. This is why their uranium
consumption is correspondingly low.

FBR's of 1000 MWe power have annual fuel consumptions as low as
approximately 1.6 te of depleted uranium or natural uranium (at
100% load factor of the plant), if they are started up with
plutonium previously generated in LWR's or in other FBR's. Initially, the converter reactors producing plutonium consume U-235
for energy generation. In LWR's usually uranium with a U-235
content of about 3% is used. This fuel must be enriched from
natural uranium which has a U-235 content of about 0.7%. The
remaining U-238 can be accumulated for the follow-on generation
of FBR's. Given the 5-15 million tonnes of natural uranium reserves now assumed to be available in the world, it is easy to
estimate that FBR's can open up an energy potential which will
be good for a few thousand years even if the annual nuclear energy generation were between 1000-2000 GWe. Besides fertile U-238,
also fertile Th-232 can be made the subject of similar considerations. With the use of FBR's, nuclear fuel supplies can therefore be considered to be inexhaustible far beyond any time scale
of conceivable planning interest. This is comparable with the
energy potential hoped to be available for harnessing in fusion
reactors, which are to operate in the D-T fusion cycle with
lithium in the breeding blanket.

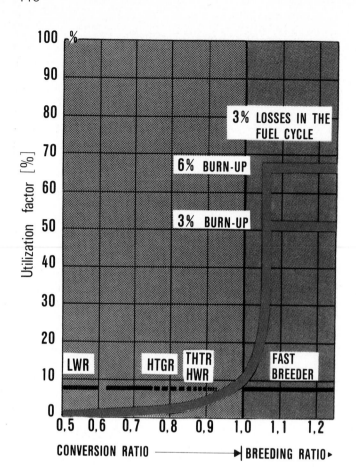

Fig. 3. Utilization of natural uranium

2 History of the Development of Fast Breeder Reactors

The breeding principle had been recognized at the beginning of the development of nuclear fission reactors. Construction of the first reactors with fast neutron spectra began in the United States before 1950; in the United Kingdom and the U.S.S.R. fast reactor development started in the early fifties. These first-generation breeder reactors, however, mainly served for studies of fast neutron physics (CLEMENTINE, EBR-I, BR-1, BR-2). In addition, they were to demonstrate the feasibility of the technical solutions adopted (EBR-II, ENRICO FERMI, DFR). Consequently, some of them still had rather low power levels. Liquid metals, such as mercury, sodium-potassium or sodium, were used as coolants at relatively moderate coolant temperatures (liquid metal cooled fast breeder reactors, LMFBR's).

Reactors up to a power of 60 MWth were built in an interim phase between 1960 and 1970. They mainly served to demonstrate the properties of ceramic fuels up to high burnups and the safety of operation of this type of reactor (SEFOR, BR-5, BOR-60, RAPSODIE). All these reactors had PuO_2/UO_2 mixed oxide fuels and were cooled with sodium. Coolant temperatures already allowed steam conditions to be chosen so as to guarantee high thermal efficiencies of these plants.

Meanwhile, breeders of this first generation have proved the good predictability of the physics characteristics of FBR cores. LMFBR cores have negative temperature and power coefficients of reactivity; the sodium components and sodium instruments have demonstrated high reliability.

The good operating experience accumulated in this first generation of test reactors has been the basis for construction of a second generation of breeder power plants with electric powers around 250-300 MWe (Table 1). The technical data and the design features of these reactors of the second generation already approach the following target data of commercial size LMFBR's: They have PuO_2/UO_2 mixed oxide fuels with target burnups of 70,000-100,000 MWd/te and sodium as a coolant. The outlet temperature of the coolant from the core is approximately 500°C, permitting steam conditions at the turbogenerator system with a thermal efficiency around 40%. The breeding ratios of these LMFBR's are around 1.15-1.20.

Regarding the primary cooling systems, two design principles are adopted: the loop-type system and the pool-type system:

In the loop-type system, the sodium coolant flows from the reactor tank through pipelines to the heat exchangers and, moved by sodium pumps, back into the reactor tank. In the pool-type system, the heat exchangers together with the reactor core and the sodium pumps are integrated in a larger tank.

Three prototype reactors in this power category by now have provided several years of operating experience. The French PHENIX prototype reactor has been in operation since 1973. The Russian BN-350 prototype breeder reactor was first connected to the grid in 1973. The British PFR prototype reactor reached criticality in 1974 and has delivered power into the public grid system since 1975.

For all three prototype plants, the original design characteristics were confirmed with respect to fast reactor core physics, control and safety engineering parameters, and the performance of the primary coolant system. The control and safety performance of these prototypes has been tested not only in normal operation, but also under simulated emergency cooling conditions and has been found to be in full agreement with expectations. Initial technical difficulties in running the large sodium components which caused some trouble in non-nuclear parts of the system (heat exchangers and steam generators) have meanwhile been overcome. PHENIX demonstrated a breeding ratio of BR = 1.16 in reactor operation; a large proportion of the irradiated fuel assemblies have already been reprocessed.

The FFTF reactor in the United States, a large test reactor of 400 MWth, was put into operation in 1979/1980 after extensive pretest programs. This test reactor is at present not used for the production of electrical power, but will mainly be used for

Table 1. Fast breeder prototype and demonstration reactors

		U.K.		France		F.R.G.	U.S.S.R.		U.S.A.	Japan
		PFR	CFR-1	Phenix	Superphenix	SNR 300	BN350	BN 600	CRBR	MONJU
Reactor power										
Thermal	MW(th)	600	2900	563	2910	736	1000	1480	950	714
Electrical	MW(e)	270(254)	1320(1250)	250	1200	312(282)	350[a]	600	360	300
Primary circuit		Pool	Pool	Pool	Pool	Loop	Loop	Pool	Loop	Loop
Number of loops						3	6 [b]		3 or 4	3
Diameter of reactor vessel	m	12.2	22.5	12.8	≃17	6.3				6.3
Coolant		Na	Na	Na	Na	Na	Na	Na	Na	Na
Coolant temperature at										
Core inlet	°C	400	400	400	395	377	300	380	387	390
Core outlet	°C	562	562[c]	560	535	546	500	550	540	540
Core dimensions										
Height	cm	91	100	85	100	95	106	75		90
Diameter	cm	136	≃270	136	≃270	157	150	205		178
Fuel		UO_2/PuO_2	UO_2/PuO_2	UO_2/PuO_2 [d]	UO_2/PuO_2	UO_2/PuO_2	UO_2	UO_2/PuO_2 [e]	UO_2/PuO_2	UO_2/PuO_2
Cladding		SS	SS	SS	SS	SS(1.4988)	SS	SS	SS 316 20% c.w.	SS
Pin diameter	mm	5.84	5.84	6.6	8.75	6.0	6.1	6.9		6.5
Number of fuel pins per fuel element		325	325	271	271	169	169	127		169
Maximum rod power	W/cm	450	450	430	450	460	470			457
Maximum clad temperature (hot spot)	°C	700	700[c]	700	680	700	680	700		700

Table 1. (continued)

Maximum fluence at clad n/cm²			$3 \cdot 10^{23}$	$4.6 \cdot 10^{23}$	2 to $3 \cdot 10^{23}$			$3 \cdot 10^{23}$	$3 \cdot 10^{23}$
Burnup MW(th)·day/t	70,000	>70,000	50,000	70,000	55,000	50,000	90,000	90,000	80,000
Breeding ratio	1.2	≈1.2	1.16	1.17	1.2		(1.4)	≥1.2	1.17
Status	In operation since 1975	Planned	In operation since 1973	Under construction, start of operation 1983	Under construction, start of operation 1985	In operation since 1973	In operation since 1980	Planned	Planned
Site	Dounreay (Scotland)		Marcoule (France)	Creys-Malville	Kalkar (F.R.G.)	Shevshenko (Caspian Sea)	Bjelojarsk (Ural)	John Sevier (Tenn.)	
Supplier	TNPG	TNPG[f]	CEA	EdF/ENEL/ SBK	INB				PNC
Operator	UKAEA	CEGB	CEA/EdF	EdF/ENEL/ SBK	SBK	State Commitee for Atomic Energy	Ministry for Electric Energy	TVA Commonwealth Edison	

[a] Or 150 MW(e) + 120,000 m³/day fresh water
[b] Including one as reserve
[c] Preliminary values, may be lower
[d] Initial loading partly with UO₂
[e] At first UO₂
[f] With SNR Consortium

fuels and materials testing. The Russian BN-600, the first FBR
with an electric power of 600 MWe, went into operation in 1980.
It represents an intermediate step between the prototypes of the
250-300 MWe class and commercial size FBR's. In the Federal Republic of Germany, a 300 MWe prototype FBR (SNR 300) is under
construction, while the small test reactor KNK-II is in operation.
Japan plans the construction of the 250 MWe prototype FBR, MONJU,
after having accumulated sufficient operating experience with
its test reactor YOYO.

The phase of commercial-size demonstration power plants was opened
up in France in 1977 with the beginning of construction of SUPER-
PHENIX. It has a net electric power of 1200 MWe and a thermal efficiency of 40%. It is scheduled to be operated at full power in
1983. In the United Kingdom, the commercial-size CDFR demonstratio
power plant is now in the detailed planning phase. A decision
about construction of this plant is expected for the next few
years. In the U.S.S.R., a 1600 MWe demonstration FBR plant has
been completed in planning, which is to be used as a prototype
for later commercial-size facilities. Similar studies for commercial-size LMFBR's are underway in the United States, the Federal
Republic of Germany, and Japan.

3 Core Design of Fast Breeder Reactors

LMFBR cores mainly consist of a cylindrical arrangement of hexagonal fuel elements surrounded radially by hexagonal blanket
elements. The hexagonal fuel elements are designed in such a way
that the inner core is also axially surrounded by an upper and
a lower breeding blanket. The blanket elements initially contain
depleted uranium as uranium dioxide (UO_2). By contrast, the fuel
elements of the core are filled with fissile material in the
form of PuO_2/UO_2 mixed oxide. Most cores contain two radial zones
of different enrichments in order to make the radial neutron flux
and power distributions as flat as possible.

In such an LMFBR core (see Fig.4a) with a net power production
of about 1300 MWe, a typical inner radial core zone contains
about 300 fuel elements with an enrichment of about 17% Pu/(Pu+U),
whereas the outer core zone usually consists of about 170 elements with roughly 24% Pu enrichment. It is surrounded by a radial blanket containing about 270 hexagonal blanket elements.
Within the core region there are roughly 25-50 positions for absorber elements containing boron carbide (B_4C) as the absorber
material. Insertion or withdrawal of these elements regulates
the criticality and the power distribution of the reactor and
guarantees safe shutdown conditions. The core has a diameter of
about 4 m, an active height of roughly 1 m, and a volume of approximately 12 m^3. This corresponds to an average power per unit
volume of the core of 280 kWth/1 and a maximum power per unit
volume of about 430 kWth/1. Blanket thickness ranges between 30
and 50 cm, and 30 and 45 cm in the axial and radial directions,
respectively.

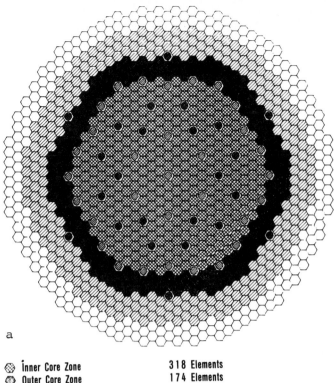

a

⊗ Inner Core Zone	318 Elements
⊗ Outer Core Zone	174 Elements
⊘ Radial Blanket	270 Elements
○ Radial Reflector	222 Elements
⊚ Control and Shutdown System	31 Elements
⬡ 2nd Shutdown System	24 Elements

Fig. 4a.
Homogeneous core design

In addition to partitioning the core radially into different
enrichment zones, there are other possibilities to obtain a
fairly flat power distribution, e.g., by inserting blanket
elements into the central fissile region. This can be done in
many different ways leading to different designs of so-called
heterogeneous cores. Figure 4b shows a typical example. This
core of uniform enrichment of about 24% contains a large central
fertile island. The other internal blanket elements are mainly
distributed on two rings, which are not completely closed. The
remaining individual internal blanket elements (fertile islands)
are positioned appropriately so as to bring about a favorable
power distribution in their neighborhood.

The fuel element is made up of 271 fuel rods kept in position by
spacers. The fuel rods have outside diameters of about 8 mm and
are fueled in the core regions with cylindrical pellets made of
PuO_2/UO_2 mixed oxide (Fig.5). In fuel rod fabrication, the rods
are finally filled with helium and welded tight at the ends. The
fuel rod cladding consists of austenitic steel of the Ti-stabili-
zed type (SS 316) with a wall thickness of approximately 0.5-0.7
mm. Below the lower axial breeding blanket there is a gas plenum
of 650-850 mm length, in which gaseous fission products accumu-

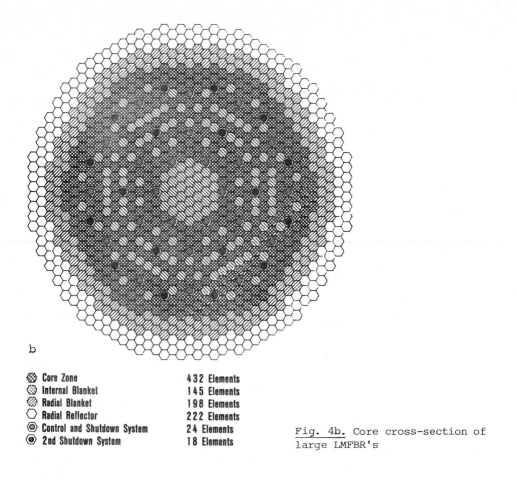

b

⊗	Core Zone	432 Elements
⊗	Internal Blanket	145 Elements
⊘	Radial Blanket	198 Elements
◯	Radial Reflector	222 Elements
◎	Control and Shutdown System	24 Elements
◉	2nd Shutdown System	18 Elements

Fig. 4b. Core cross-section of large LMFBR's

late. Sodium flows through the fuel elements at a rate of approximately 3-6 m/s. On its way through the fuel elements, the sodium is heated from 390°C to 540°C. The assembly of fuel rods is inserted into a hexagonal box made of austenitic or ferritic steel of a wall thickness of 4.5 mm. The whole fuel element has a length ranging 4.2-5.4 m for different designs and is typically 16 cm across the flats. The blanket rods usually have diameters of 12-16 mm. They can be thicker than fuel rods because the power production per unit volume of fuel in the blanket is considerably smaller than in the core. Blanket elements of the same size as the core elements contain typically 91 to 127 blanket rods. The maximum power of the fuel rod in the center of the core varies for different core designs between 415 and 450 W/cm of fuel rod length. A 1300 MWe core contains some 5 te of Pu_{fiss} corresponding to 3.9 te Pu_{fiss}/GWe. The core contains a total of 36 te of PuO_2/UO_2, the axial breeding blanket of 50 cm thickness at the top and the bottom of the core contains 40 te of UO_2, and the radial blanket roughly 50 te of UO_2. Absorber elements contain about 20-40 absorber rods with outside diameters of 17-19 mm, which are filled with boron carbide (B_4C) pellets. In order to enhance the effect of the absorber elements, the

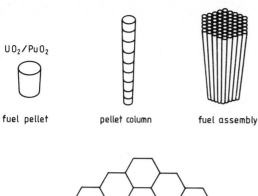

Fig. 5.
Breeder-fuel elements

UO₂/PuO₂

fuel pellet   pellet column   fuel assembly

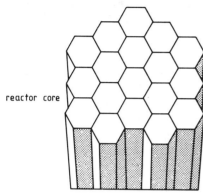

reactor core

concentration of the B-10 isotope may be increased up to 93%
whereas, in natural boron, the B-10 isotope abundance is roughly
20%.

4 Technical Aspects of Sodium-Cooled FBR's

Sodium as a coolant of fast reactor cores at present clearly
dominates in fast reactor development and demonstration programs
all over the world. Helium cooled fast reactor design concepts
were pursued as an alternative coolant concept. However, no test
or demonstration fast reactor on the basis of helium as a coolant
has been constructed so far.

The design concept of liquid sodium-cooled fast breeder reactors
is mainly determined by the choice of sodium as a coolant. Sodium is chosen because of its good heat-transfer properties and
its small moderating effect. Sodium has a small neutron capture
cross section resulting in the formation of Na-24 with a half-
life of 15 h. This makes the primary sodium radioactive on its
passage through the core. Sodium has a high melting point (98°C)
and a high boiling point well above the core outlet temperature
(892°C at atmospheric pressure, and 900-1000°C at coolant pressures within the core). It has a high specific heat and very
good thermal conductivity.

The high melting point of sodium requires preheating systems of
pipes and components of the cooling systems before operational
conditions are reached at start-up. The high boiling point allows

for high coolant temperature conditions under very low system
pressures (6 to 10 bar). This results in high thermal efficiencies
for the overall LMFBR plant in the range of 40%. The relatively
high specific heat permits moderate coolant velocities of 2 to
6 m/s within the fuel elements and low pumping power, whereas
the good thermal conductivity together with other thermal properties lead to extremely good natural convection conditions in the
core and coolant system during shutdown and emergency core cooling conditions. However, these excellent thermal properties also
give rise to special design and operational consequences. Thermo-
shock problems within the reactor vessel at pipe nozzles and
valves must be avoided during short-term power reductions or reactor scram conditions.

Although sodium is not very corrosive against stainless steel,
its impurities, mainly oxygen and carbon, must be held at very
low contents by special cold traps in the bypass of the coolant
systems (5 to 10 ppm for O_2, 50 ppm for carbon). High impurity
contents cause radioactive corrosion products to dissolve from
the surfaces of the fuel claddings and to be carried to low-
temperature parts of the primary coolant system (heat exchangers).
This must be prevented to avoid maintenance and repair difficulties, which could arise after several years of plant operation.

The opacity of sodium affects the design of the refueling systems and sometimes requires ultrasonic devices for the super-
vision of refueling and repair processes.

Major design consequences arise from the potential of sodium to
start chemical reactions with water and air. This property together with the fact that sodium becomes radioactive under neutron irradiation in the core leads to a plant design with

   a primary coolant system containing the radioactive sodium
   heated up in the core;

   a secondary coolant system coupled with the primary one by
   intermediate heat exchangers;

   a tertiary water system producing steam for electricity
   generation by means of the turbogenerator system (see Fig.6).

Within the primary coolant system, radioactive sodium is protected from air by steel barriers and cells filled with a cover
gas, usually argon or nitrogen. Radioactive sodium of the primary coolant system is separated from the non-radioactive sodium of the secondary system by the steel tubes of the intermediate heat exchangers.

So far, two principal design concepts have been used for LMFBR's,
the pool- and the loop-type concepts (Fig.7). In the pool-type
concept, the primary system components together with the core,
primary pumps, and intermediate heat exchangers are built into
the sodium-filled pool tank. This concept was used in PFR,
CFR-1, PHENIX, SUPERPHENIX and BN 600.

In the loop-type concept, by contrast, only the reactor core is
built into the reactor vessel, and the primary sodium is pumped
to the intermediate heat exchanger via a piping system. This

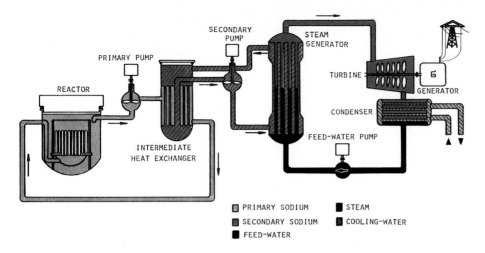

Fig. 6. Sodium and steam circuits of loop-type LMFBR's

design concept was chosen for BN 350, SNR 300, FFTF and MONJU. Both design concepts have a number of advantages and disadvantages which roughly balance out. Only future experience with operation, maintenance, and licensing of large LMFBR's can prove whether one of these design concepts offers in the long range some economic or performance advantages.

Loading and unloading of fuel, radial blanket and absorber elements are done with special refueling machines placed at the top of the reactor vessel. All LMFBR's, therefore, have rotating plug systems at the tops of their reactor vessels. These rotating shield plug systems consist of eccentric arrangements of up to three smaller rotating plugs which can be moved against each other. Rotating plug systems allow all positions of the different fuel elements in the honeycomb structure core to be reached.

As with the fast reactor core, much care is required in designing and building of sodium-heated steam generators. LMFBR steam generators contain non-radioactive secondary-system sodium and water separated from each other only by the tube walls. Because of their different physical and chemical properties, sodium and water are able to initiate a violent chemical reaction, if appreciable amounts got into contact with each other. Many design aspects have to be taken into account, such as fabrication, safety, operational availability, leak detection, inspectability, corrosion effects, repair, etc. Therefore, much research within the LMFBR programs has been devoted to steam generator development and testing in full-scale test facilities. Figure 8 exemplifies two different concepts of steam generators, a straight tube steam generator and a helical tube steam generator, both of the SNR 300 plant. A number of other design concepts are used by other fast reactor projects. Sodium-water interactions occur if one of the water-filled tubes in a steam generator fails. Water or steam is then released under high pressure into the

Pool Reactor

Fig. 7. Alternative LMFBR design concepts - pool and loop

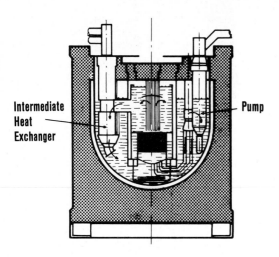

Intermediate Heat Exchanger

Pump

Loop Reactor

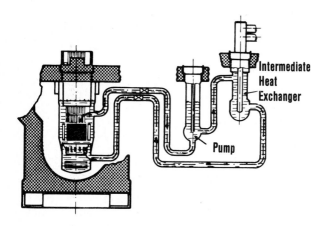

Intermediate Heat Exchanger

Pump

sodium. Sodium-water interaction, hydrogen production, and pressure generation result. A special pressure relief system using rupture disks then releases hydrogen into the atmosphere. Since the secondary-system sodium does not become radioactive, no radioactivity is released into the atmosphere during a sodium-water interaction in a steam generator.

Much research has been carried out in the past 10 years to investigate these sodium-water interactions and optimize the designs of pressure relief systems. There is no doubt that large LMFBR steam generators can be based upon a sound and safe technological concept. Analyses of the difficulties encountered with steam generators in the early phases of BN 350 and PFR operation have shown that there would be no major technological problems with LMFBR steam generators. However, great care must be devoted

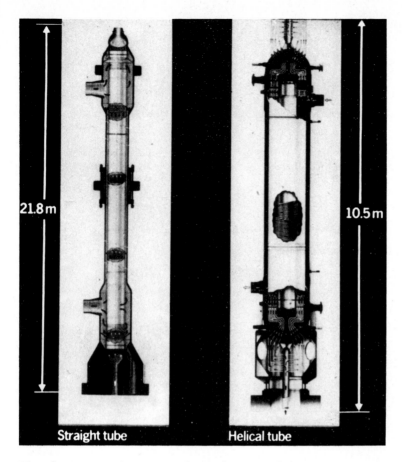

Fig. 8. Steam generators of SNR-300 fast breeder prototype power plant

21.8 m — Straight tube

10.5 m — Helical tube

to the proper choice of tube steel (intercrystalline corrosion, weldability, etc.) and to quality assurance during fabrication, if the expected high plant availability of future large commercial size LMFBR's is to be reached.

5 Safety Design Aspects of LMFBR Plants

As in all nuclear reactors, the objective of all safety design measures in LMFBR's is to prevent the radioactivity present in the reactor core from ever penetrating into the environment of the plant, either in normal operation or under accident conditions. LMFBR plants produce roughly the same amounts of fission products per GWth·a as converter reactors. The fission product isotopes generated differ only slightly from the fission product distribution encountered in converter reactors. The quantities of higher actinides produced in LMFBR's per GWth·a are slightly larger than those in converter reactors. However, the absolute quantity of fissile plutonium differs relative to U-235 and U-233 fueled converter reactors. The total fuel inventory

in the core and the blankets, which is about 120 te of fuel at
1300 MWe, is comparable, e.g., with 104 te of UO_2 in a 1230 MWe
PWR.

5.1 The Multiple Barrier Principle

One of the most important safety design principles applied in
LMFBR's, as in any other reactor concept, is the principle of
multiple safety barriers between the fuel generating energy and
thus also radioactive fission products, and the environment.
This principle will be explained by the example of the SNR 300,
a loop type reactor (Fig.9).

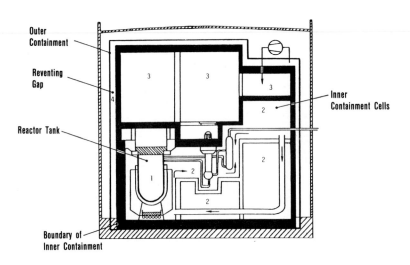

Fig. 9. Containment design of loop-type reactor SNR-300

The first solid barrier against the fuel is the tightly welded
steel cladding of the fuel rod. Solid fission products remain in
the crystalline structure of the fuel. Gaseous fission products
are collected in the fission product gas plenum of the fuel rod.
Leaks in the cladding are detected by means of fission product
gas monitors in the cover gas or by means of delayed neutron
monitors in the primary sodium. If necessary, the reactor will
be shut down so that the element containing the defective rod
can be replaced.

The second barrier is constituted by the reactor double tank and
the primary system. The reactor tank is limited at the top by
the reactor head. Argon as a cover gas is kept between the free
sodium level and the tank head. The intermediate heat exchangers
(IHX's) and the pumps of the primary systems are installed in
so-called inner containment cells (third barrier). The interior
of these inner containment cells is lined with steel plate and
filled with nitrogen. The pipings, pumps, and IHX's are installed
in such a way that, in case of a leak in the primary pipelines,

the primary sodium will flow into special cavities and a certain sodium level in the reactor tank will not be underrun. In LMFBR's, in which the systems are only kept under pump pressure to overcome flow losses, sodium leaks cannot easily result in large amounts of sodium flowing out. Sodium leaks can also be detected immediately by a number of sodium leak detectors.

The fourth barrier is constituted by the outer containment filled with air. In the SNR 300, this outer containment consists of reinforced concrete with a wall thickness of approximately 1.2 m, protecting the plant against such external impacts as airplane crashes, earthquakes, etc. It is completely surrounded by a steel liner and has a reventing gap between the concrete walls and the liner. Any radioactive aerosols leaking into this gap can be revented by blowers into the outer containment. This reventing procedure can be kept up for a period of days, if needed. After this time period, aerosols can be released through filters to the stack.

5.2 Control and Shutdown Systems

LMFBR cores have a negative overall temperature coefficient and a negative power coefficient. This makes them inherently stable for control purposes. They are controlled by means of absorber rods containing B_4C as an absorber material. Also shutdown can be brought about by means of B_4C absorber rods. After being released, the absorber rods can enter into the core by dropping under gravity within 0.7-0.8 s. To add to the reliability of this shutdown concept, e.g., the SNR 300 as well as other FBR's have two completely independent and diverse shutdown systems. In this way, a failure rate for the shutdown systems of $<10^{-6}$ failures/a can be attained. By way of example, Fig.10 shows the design principle of two independent shutdown systems in the SNR 300. The primary shutdown system drops absorber rods into the core. The secondary shutdown systems pulls a flexible absorber chain into the core from below. Both systems have diverse electronic channels. The magnetic release of the absorbers is direct and indirect, respectively. In SUPERPHENIX, some of the absorber rods are additionally released by inherent mechanisms as soon as the sodium temperature in the core rises.

5.3 Afterheat Removal and Emergency Cooling of LMFBR Cores

Even after shutdown of the reactor power, the afterheat of the LMFBR core must still be safely removed. Under normal conditions it is carried to the steam generators by the main heat systems in such a way that the pumps are driven by pony motors at low speed. However, sodium has such excellent natural circulation characteristics that the afterheat can also be dissipated from the reactor core through the main systems by means of natural circulation. This has been proven experimentally both for pool-type reactors, such as PHENIX and PFR, and for loop-type reactors in SEFOR. In case the main heat transfer systems are not available, SUPERPHENIX has additional systems, which will then

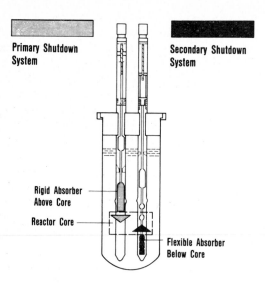

Primary Shutdown System

Secondary Shutdown System

Rigid Absorber Above Core

Reactor Core

Flexible Absorber Below Core

Fig. 10. Primary and secondary shutdown system

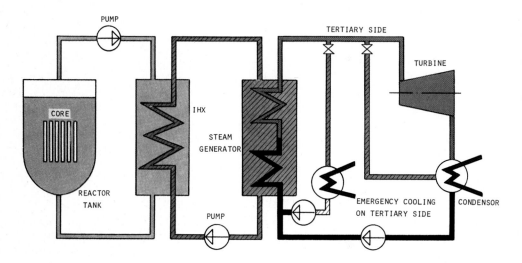

Fig. 11. Emergency cooling system for SNR-300

be initiated to transport the heat from the secondary system to sodium-air coolers. In the SNR 300, six immersion coolers arranged in the reactor tank below the emergency sodium level transfer the afterheat to sodium-air coolers (Fig.11). Even if all active components of the afterheat removal systems were to fail, enough heat would still be delivered through the surfaces of the pipelines of the primary system in the SNR 300 to keep the maximum temperature in reactor tank at <700°C. LMFBR's have in principle the inherent potential to dissipate afterheat by means of natural convection without any active systems.

5.4 Core Instrumentation and Protection Against Fault Propagation

In addition to the required high reliability of the safety shut-down systems it must also be assured that the initiation of faults — which could develop from local blockages in fuel elements — is counteracted from the beginning. Local blockages followed by local boiling and fuel pin failure were considered to have the potential to lead to partial fuel element destruction and, possibly, more severe consequences to the core. Although recent in-pile R&D programs indicate that such accident developments are not possible, the cores of LMFBR's are equipped with individual subassembly instruments (thermocouples and delayed neutron monitoring) to detect coolant anomalies and fuel rod failures, and shut the reactor down.

Other possible faults in loop-type reactors are counteracted by inherent design measures or additional detectors. The SNR 300 is equipped with a gas bubble separator underneath the core diagrid plate to prevent any gas from passing through the core. Piping integrity is surveyed by in-service inspection methods, sodium level detection in the reactor vessel and expansion tank, and by leak detection systems.

5.5 Safety Design Basis of the Primary System and Containment

While, in LWR's, the loss-of-coolant accident (LOCA) must be regarded as a design-base accident, it is a different class of extremely improbable accidents which play a major role in the safety analysis of LMFBR's. In LMFBR's, core destruction can occur only if the pumps are shut down after scram initiation while, at the same time, the two shutdown systems fail. In this so-called unprotected loss-of-flow (LOF) accident, the reactor power remains constant, while the coolant pumps and thus sodium cooling will be shut down within in a few seconds. Accidents initiated by positive reactivities are less probable, but have very similar consequences. Failures of the two redundant and diverse shutdown systems are supposed to occur only with a probability of $<10^{-6}$/a. Accordingly, also the unprotected LOF accident must be classified in this region.

Safety analysis of the unprotected LOF accident in the SNR 300 shows that there will be a rapid temperature rise in the sodium coolant in the reactor core because of the imbalance between power generation and cooling. When the boiling temperature of sodium will have been reached, which is 900-1000°C, sodium will begin to boil in the core. Initially, because of the positive sodium void coefficient, there will be further power increases, which are counteracted by negative temperature and power coefficients, e.g., the Doppler coefficient and the fuel expansion coefficient. However, the temperatures reached will be high enough for steel, fission products and PuO_2/UO_2 mixed oxide fuel to melt, boil and eventually evaporate in the center of the core. For a few milliseconds, this hot mixture of core material will be under a pressure of a few tens of atmospheres locally and will expand rapidly. This will make the reactor core subcritical and cause it to shut itself down automatically. In the SNR 300 core, the isentropic work potential of this mixture of hot core material will be below 100 MJ. This is well below the design load limit of 370 MJ for the mechanical integrity of the reactor tank system.

A large part of the LMFBR core will have molten after this accident. Cooling of this molton core debris must be ensured also for this extremely unlikely case. For the SNR 300, theoretical and experimental analyses show that cooling in the reactor tank is ensured and molten fuel cannot melt through the lower tank bottom. For other demonstration LMFBR's and in SUPERPHENIX similar conclusions were reached. In the SNR 300, however, an additional safety barrier was introduced for the unlikely event that molten fuel were to melt through the tank bottom.

In case the second containment barrier were to develop a leak, the third and fourth barriers will still retain their sealing functions against radioactivity releases even in this severe accident under discussion. Sources of radioactivity to be considered will primarily be gaseous and volatile fission products as well as sodium aerosols. In loop-type reactors, such as the SNR 300, the inner containment cells would represent the third barrier, the outer containment with its reventing capability the fourth barrier. The retention capability of the containment (barriers) and the filter systems is designed in accordance with aerosol physics. Most of the aerosols entering the spaces in the third and fourth barriers would plate out on the walls and at the bottom by coagulation and sedimentation processes. Only plutonium aerosol concentrations in the mg/m^3 range would be found to remain airborne in the outer containment of the SNR 300. After a few days, it would be necessary to release some of the contents of the outer containment over sandbed and charcoal filters through a stack into the environment. The resulting irradiation dose at the fence of the reactor plant could remain below the legally permissible limits.

5.6 Sodium Fires

Sodium fires in the primary cooling systems with radioactive sodium are prevented by enclosing these systems in nitrogen-filled cells. Free sodium surfaces in tanks are covered with argon. Sodium leak detection systems survey the leaktightness of the cooling system. The secondary systems are usually surrounded by air.

Sodium leaking out of the cooling systems will be directly collected in special catch pan systems and containers underneath the piping systems and pumps. Access of oxygen to the hot sodium will largely be prevented in those systems. Reliable fire extinguishing systems are also available. Extensive experimental data on sodium burning rates, burning temperatures, and sodium aerosol formation are at hand from large out-of-pile test rigs in the United States and in Europe. In addition, extensive operation of demonstration LMFBR's has proved the high standard of experience of sodium technology.

6 The Fast Reactor Fuel Cycle

FBR's can only be operated in a closed fuel cycle (Fig.12). The necessary reprocessing and refabrication technologies therefore must be developed along with reactor technology. The PUREX process well known from reprocessing of LWR fuel is also being applied in chemical reprocessing of irradiated PuO_2/UO_2 FBR fuel. Special properties of irradiated FBR fuel, e.g., the relatively high plutonium enrichment and high burnup, must be taken into account by appropriate process modifications.

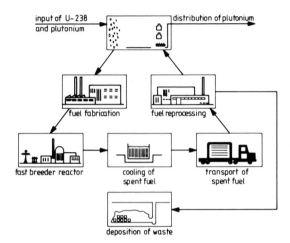

Fig. 12. Fast breeder reactor fuel cycle

In the United Kingdom and France, small pilot reprocessing and refabrication facilities are already in operation with annual capacities of 5-10 te of fuel. As more commerical size FBR's will come into operation, also the size of reprocessing facilities must be increased. Commercial size reprocessing and refabrication plants will be able eventually to serve 10-15 GWe in FBR plants.

Acknowledgements. The author wishes to thank his colleagues D. Faude and Dr. E. Kiefhaber for their assistance and advice in writing this paper.

References

Häfele W, Holdren JP, Kessler G, Kulcinski GL (1977) Fusion and fast breeder reactors. Internatl. Inst Appl Syst Anal (IIASA). Laxenburg, Austria, RR-77-8

Internatl Conf Engineering of fast reactor safe and reliable operation. Karlsruhe, October 1972 Ges Kernforsch Karlsruhe

Int Meet fast reactor safety technology. Seattle, August 1979. Am Nucl Soc, Eur Nucl Soc

Internatl Nuclear Fuel Cycle Evaluation (INFCE) (1980) Fast breeders, Rep Working Group 5. IAEA, Vienna

Internatl Symp design, construction and operating experience of demonstration LMFBR's. Bologna, April 1978 IAEA, Vienna, SM-225

Yevick JG, Amorosi A (1966) Fast reactor technology. MIT Press, Cambridge, Mass

Fast Breeder Reactors in France in 1979

M. Rapin[1]

1 The French Energy Situation

The French energy situation can easily be characterized by three
figures:

    3   % of the world energy consumption
    0.7 % of the world energy production
    0.11 % of the world proven energy reserves

Domestic energy reserves are largely unbalanced compared to our
own needs.

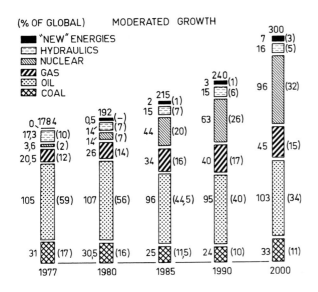

Fig. 1. Contribution of the various sources to the primary energy French consumption in MTOE

The French total primary energy consumption was about 182 MTOE
in 1978 (Fig.1). Even assuming a limited economic growth rate
compared to the past, and taking into account the possibility to
reduce the elasticity of the energy demand versus economic growth,

1 Commissariat à l'Energie atomique, Directeur Délégué aux Applications Ener-
  giques Nucléaires, 31-33 Rue de la Fédération, 75732 Paris Cedex 15, France

situations of the various countries. However, I think that time is pressing for everybody, taking into account the inertia of any energy system.

In any case, in a country like France, considering our energy situation and our uranium reserves, the question does not even arise. And it is only logical that our government should be firmly decided to push ahead with the breeder reactor programs.

2.2   FBR French Program

LMFBRs studies have been folloging in France, as in the majority of the other countries engaged in that way, a logical progression for 20 years.

After an initial phase aiming at studying core physics and sodium and fuel technologies, experimental reactors were built to settle the validity of the concept and to perform the irradiations of the fuel. In France, RAPSODIE, operational at Cadarache in January 1967 reached a power of 40 MW thermal in 1970 in the second version RAPSODIE-FORTISSIMO. It is a remarkable test stand for components, instrumentation, fuels, materials, and reactor technology as a whole.

The industrial phase of the LMFBR development starts really with the demonstration reactors. In France, PHENIX, a 250 MWe plant, in industrial operation at MARCOULE since 1974, has produced till today more than 6 billions KWh (Fig.10). Its average load factor during these 5 years was only 54% because of the long shutdown necessary for the repair of the intermediate heat exchangers. From April 1979 on, date of the restart-up at full power, its load factor exceeded 75% and reached 82% during the first half of 1979.

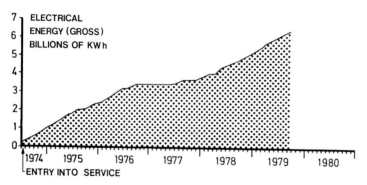

Fig. 10. Phenix nuclear plant electrical energy production

The whole experience gained from the satisfactory operation of PHENIX led the company NERSA, created by three large European electricity suppliers, to launch in April 1977 the first unit of quasi-industrial size, CREYS-MALVILLE, a 1200 MWe plant, based on the SUPERPHENIX 1 boiler built by an association of the companies NOVATOME (France) and NIRA (Italy). The construction of

Fig. 11. The Creys-Malville plant

this plant on the RHONE river is progressing satisfactorily, in close accordance with the schedule which plans industrial service in 1983 (Fig.11).

For the future, a new boiler, SUPERPHENIX 2, with a power of the order of 1500 MWe, is now being studied by NOVATOME for EDF with the main goal of reducing investment costs as compared to SUPERPHENIX 1.

2.3 Fuel Cycle Development

In parallel to reactor construction, the various steps of the LMFBR fuel cycle, especially fuel fabrication, reprocessing and wastes aspects, were thoroughly studied inside the CEA group.

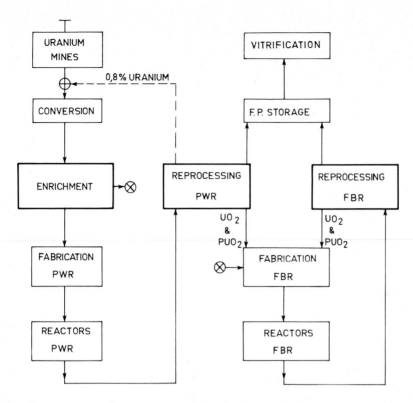

<u>Fig. 12.</u> Uranium-plutonium cycle — transient phase. Depleted uranium coming
from the enrichment operations. *NB* The material mass flows corresponding to
the U-PU cycle in the transient phase depend on the relative percentage of
PWRs and *FBRs*

It is well known that, for a transient phase, the LMFBRs will
mainly be fed by plutonium coming from PWRs (Fig.12.).

The present growth of the nuclear program, based today on PWRs,
implies a simultaneous development of all the steps of the fuel
cycle industries. I will just cover briefly the main aspects of
the thermal fuel cycle works related to the FBRs.

Even with the planned FBR program, the natural uranium needs for
the PWRs ask for significant prospection efforts in France and
all over the world.

The introduction of FBRs significantly decreases the enrichment
needs: for example from 10 million SWU a year for PWRs alone
down to 6 million SWU a year for a symbiosis PWRs - FBRs in 2010
(Fig.13). Up to the 90s, the French share in the EURODIF multi-
national plant, at TRICASTIN, with a nominal annual capacity of
about 10 million SWU will cover our needs. EURODIF is now operating with a capacity of the order of 2,6 million SWU a year.
May I mention, at this time, that the total amount of depleted
uranium which will be produced till the turn of the century at
EURODIF will reach about 250,000 t. If there were no breeder

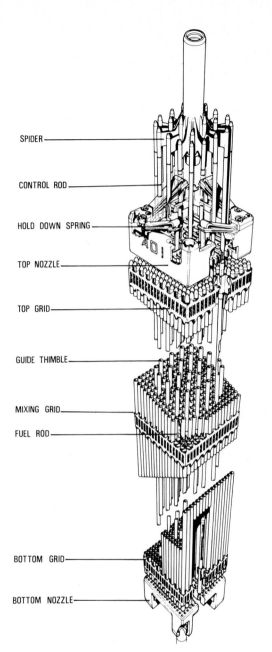

Fig. 4.
17 × 17 fuel assembly

SPIDER

CONTROL ROD

HOLD DOWN SPRING

TOP NOZZLE

TOP GRID

GUIDE THIMBLE

MIXING GRID

FUEL ROD

BOTTOM GRID

BOTTOM NOZZLE

pellets (UO_2) are located at both ends of the fissile column to achieve the breeding properties.

Due to the fuel characteristics (use of plutonium enrichments in fissile material much higher than for LWR: 15%-25% instead of 3%) particular precautions have to be taken during the fabrication:

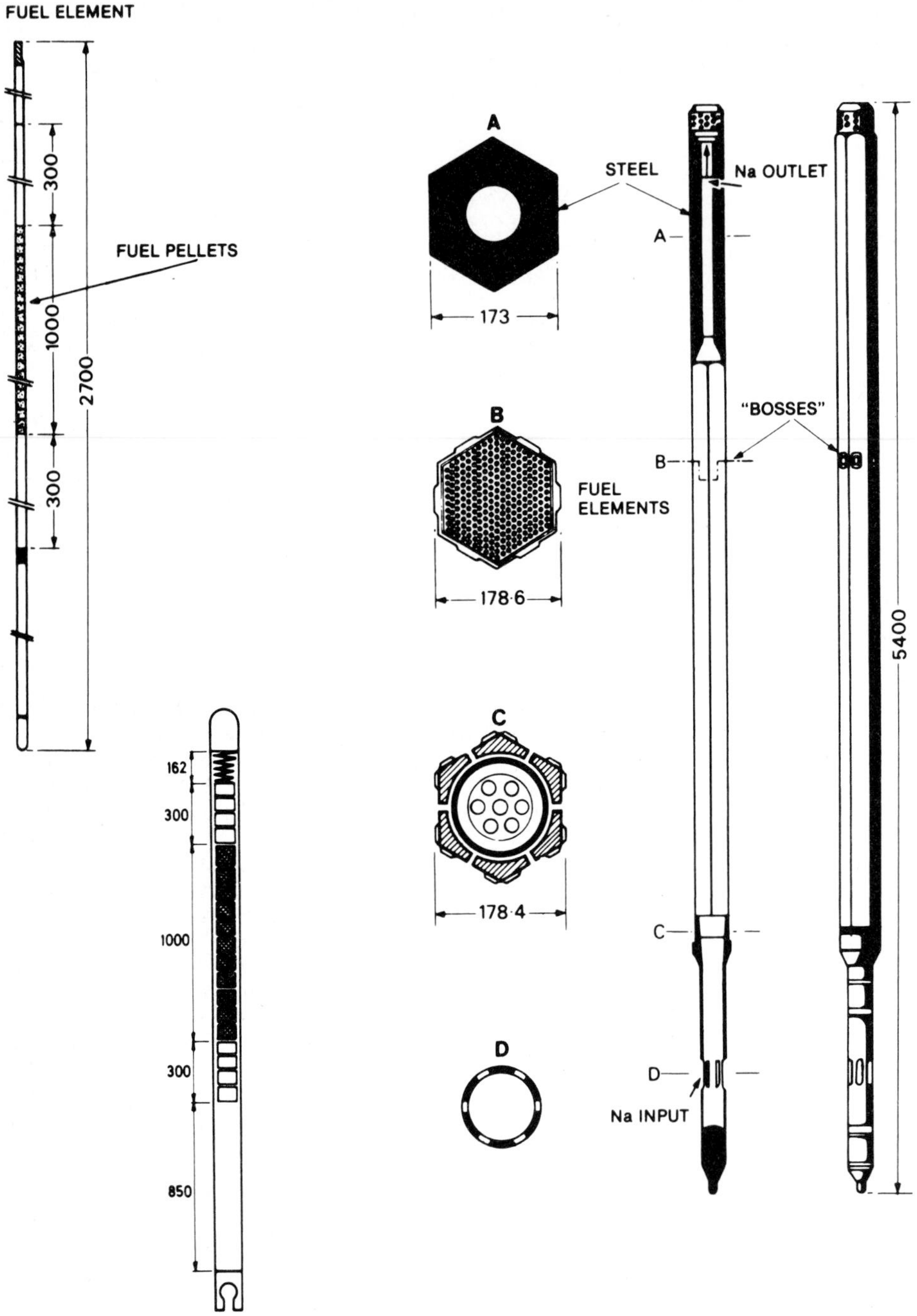

FUEL ELEMENT

FUEL PELLETS

300

1000

300

2700

A

STEEL

Na OUTLET

173

A

B

FUEL
ELEMENTS

178·6

"BOSSES"

B

C

178·4

C

D

D

Na INPUT

5400

162

300

1000

300

850

Fig. 5. Fuel assembly

plutonium isotope confinement (glove, boxes) against α emission

shielding against γ and neutron emission,

crisis prevention,

no possibility of fuel diversion.

Up to now, in most cases fuel fabrication experience has been gained with test and demonstration reactor cores in relatively small capacity plants:

	Approximate heavy metal capacity t/year	Fuel fabricated for
United States	5	FFTF
Belgium	5	SNR 300
Great Britain	5	PFR
Japan	10	JOYO
Federal Republic of Germany	10	KNK II SNR 300
France	20	RAPSODIE PHENIX SUPER-PHENIX

The first experience related to large-scale fabrication is in progress in France with the SUPER-PHENIX core-loading fabrication, the characteristics of which are listed hereafter:

	First loading	Annual reloadings
Bundles	400	180
Pins	120,000	54,000
Core heavy metals	$\simeq$ 36 t	16 t
Pu	5.6 t	2.5 t
Axial blanket (depleted U)	25 t	11 t
Fertile subassembly	240	60
Fertile pins	24,000	6,000
Radial blanket (depleted U)	52 t	14 t

Insofar as one proceeds from the workshop size to an industrial size plant (approximate capacity: 100 t/year) the following problems have to be solved:

increasing capacity up to $\simeq$100 t/year implies transition from manual work, which is acceptable under specified conditions (contamination, radiation exposure), to automatic devices with related reliability requirement.

160

Such reliable confinements allow to store active products without contamination problems for a long period in intermediate storage, thereby affording time to find the optimal final storage.

From the non-proliferation point of view, reprocessing allows:

to concentrate all the spent fuel in a few reprocessing sites more easily controlled than disseminated storing pools,

to store Pu in reprocessing sites, possibly in international storages, easy to manage,

to burn Pu in FBRs, which is the best way to prevent its diversion.

Concerning the fissile isotopes, reprocessing allows to recycle [^{235}U] in LWRs, which leads to a saving of 20% of the natural uranium consumption and 4% of S.W.U., and [^{239}Pu] mixed with depleted U in FBRs, which can thereby provide a long-term solution to the energy problem (Fig.7).

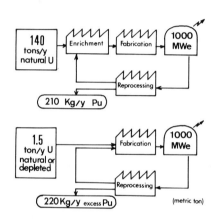

Fig. 7. Light water reactor (above) and fast breeder (below)

5.1 Thermal Reactor Fuel Reprocessing

All the major reprocessing plants in operation or in design use the PUREX process, which includes the following main steps:

fuel subassembly dismantling, e.g., by shearing and dissolution in nitric acid,

separation of uranium and plutonium from fission products and actinide isotopes by solvent extraction,

transformation of recovered uranium and plutonium in usable solid materials (plutonium oxide, uranium nitrate),

waste conditioning.

The final aim is to achieve high efficiency for fissile element recovery (>99,5%) with quite high decontamination ratios ($\approx 10^7$), which implies several consecutive extraction cycles followed by one or two purification cycles.

Important R and D work is in progress, especially in France, in order to optimize all the reprocessing steps both from the economic and technical points of view for the plants under construction or in design.

Experience on thermal fuel reprocessing has been gained up to now mainly in the Federal Republic of Germany, the United Kingdom, Japan, France. It must be underlined that if future units will deal with oxide fuel issuing from LWR plants, an important part of past experience has been obtained on other fuels (e.g., metallic uranium) used in other reactor types: NUGG (France), or AGR (United Kingdom), or MZFR (Federal Repulic of Germany).

United Kingdom: BNFL's reprocessing is performed at Windscale:

> metallic uranium fuel since 1952: more than 20,000 t of magnor fuel has been reprocessed,
>
> oxide fuel (less than 100 t) mostly issued by foreign LWRs has been reprocessed between 1969 and 1973.

In order to reprocess LWR fuel on an industrial scale, BNFL has launched the construction of a new unit at Windscale: THORP, which will have a 1300 t/year capacity and should start reprocessing by the early 90s.

Federal Republic of Germany:
> German reprocessing experience was gained in the WAK pilot plant, at Karlsruhe, with a 35 t/year nominal capacity. In this plant, 114 t of uranium oxide coming from LWRs ($\approx$50 t) and from MZFR ($\approx$60 t) have been successfully reprocessed. A next step, presently under investigation, should be reached with a 350 t/year-capacity plant which should be fully operational by the early 90s.

Japan:
> The Tokai-Mura plant, with a nominal capacity of 200 t/year started in September 1977. About 80 tons of oxide fuel have been reprocessed up to the end of 1980. An industrial size plant (1500 t/year capacity) is presently under study and is expected to start by the 90s, provided that corresponding agreements are concluded with the United States.

France: Past experience of Cogema concerns metallic uranium of the UGG plants, and oxide fuel issued by LWRs. The NUGG fuel from five reactors representing 2300 MWe is presently reprocessed in the UP 1 unit now under extension at Marcoule. What concerns LWR fuel, the large French light-water reactor program, which aims at delivering 40 GWe in 1985 and about 60 GWe in 1990, leads to a rapid increase of the cumulated quantities of irradiated fuel unloaded from reactors (Fig.8):

The corresponding annual reprocessing needs will reach 700 t/year in 1985 and $\approx$1300 t/year in 1990.

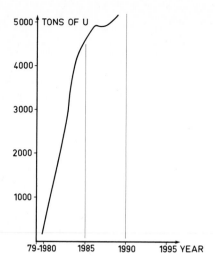

Fig. 8. Irradiated fuels waiting for reprocessing and stored in pools

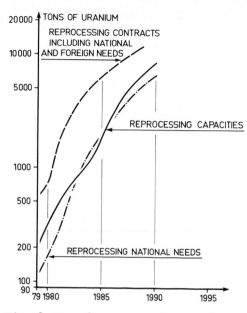

Fig. 9. French reprocessing cumulated capacities and needs

The LWR fuel reprocessing technique, intensively studied within the CEA Group, has been tested and improved first in the UP 2 unit, located at La Hague, initially devoted to NUGG irradiated fuel reprocessing, and now extended to LWR fuel by building the HAO workshop (High Activity Oxide), which started in 1976. Since that year, 300 tons of LWR fuel with high burnup (20% higher than 30,000 MWd/t, 7% higher than 33,000 MWd/t) have been reprocessed in one plant up to now. Among these 250 t, 140 were reprocessed in one single campaign betwen November 1979 and June 1980: this shows the satisfactory availability of the UP 2 unit.

In order to meet the national reprocessing needs and fulfill its contracts with foreign countries within the coming 15 years, Cogema has started the extension of the UP 2 unit to reach a 800 t/year capacity and launched the construction of a new 800 t/year capacity unit, UP 3 (Fig.9).

5.2 FBR Fuel Reprocessing

With respect to LWR fuel reprocessing, FBR fuel reprocessing, although it uses the same PUREX process, presents some differences due to FBR spent fuel characteristics:

   presence of residual sodium,

   stainless steel of pin cladding and wrapper,

   high residual power and radioactivity level due to shorter cooling time (6 months to 1 year instead of ≈3 years),

INITIAL COMPOSITION REACTORS	U/U+Pu% (CORE)	ATI (LA HAGUE) Qty(kg) U+Pu	GWd/te OXIDE CORE	SAP (MARCOULE) Qty(kg) U+Pu	GWd/te OXIDE CORE
RAPSODIE (1st CORE)	25	250	40-55		
RAPSODIE FORTISSIMO	30	658	50-120	50	14-76
PHENIX ENRICHED U	Ue			2600	38-45
PHENIX Pu CORE 1	18	177	8-44	150	37
PHENIX Pu CORE 2	25			1060	36-51

PHENIX Pu CORE 1 REPROCESSED AT UP2 (HAO) PLANT LA HAGUE	QUANTITY (kg): 2120 GWd/te OXIDE: 21-42

Fig. 10. French fast fuel reprocessing experience

high plutonium content ($\approx$10 to 15% instead of 1%),

high fission product content (10% instead of 4%).

Past experience has been obtained, up to now; mainly in the United Kingdom and France, in workshops with limited capacities, with fuel irradiated either in test reactors (e.g., RAPSODIE) or demo plants (PFR, PHENIX):

at Dounreay, 1.25 t of heavy materials were reprocessed in 1980: the corresponding fuel subassemblies were burnt up to 6,5% and stored 1 year for cooling. The PFR fuel cycle should be closed in a quite near future

in France (Fig.10), the AT1 workshop at La Hague with a nominal capacity of 1 kg per day, has reprocessed about 1 ton of RAPSODIE fuel with burnup ranging between 40,000 MWd/t and 120,000 MWd/t: the RAPSODIE cycle has been carried out several times. AT1 has also reprocessed 1980 kg of PHENIX fuel on an experimental basis. The pilot plant SAP at Marcoule (capacity between 10 and 20 kg:day depending on the Pu content) has reprocessed about 6 tonnes of FBR fuel coming mainly from PHENIX. A part of this reprocessed fuel allowed to close the PHENIX cycle within 1 year, from spent fuel unloading to the loading of a new fuel fabricated with the recovered Pu. It must be also noted that 2.1 tonnes of PHENIX fuel have been reprocessed by dilution with gas-graphite reactor fuel in La Hague facilities.

In order to complete and improve existing means, the TOR project has been launched, with a view to bringing the capacity of the Marcoule pilot plant (2.3 t/year in 1981) to about 5 t/year, thus enabling the reprocessing of PHENIX fuel and providing the possibility of reprocessing fuel coming from other fast breeders (KNK II, SNR) and eventually, as a demonstration, some subassemblies of SUPER-PHENIX 1. The TOR project should start operating by 1983. Industrial-size plant designs, which are presently under study in the United Kingdom, the United States, and France, concern 50 t/year, 100 t/year, and about 120 t/year capacities, respectively, and large-scale R and D programs are carried out in these countries to support the designs.

In a longer-term perspective the fluoxide process is also investigated in several countries.

6 Waste Management

Radioactive wastes are produced at each step of the nuclear fuel cycle and must be conditioned and managed, mainly according to their radioactive characteristics, to ensure safe and reliable long-term storage for each kind of waste (Fig.6):

low- and middle-activity level wastes containing short period isotopes and negligible amount of long periods ones (mainly α emitters),

low- and middle-activity level wastes containing non-negligible quantities of long period elements,

high-level wastes including fission products and actinides (α emitters) issued from reprocessing, and vitrified).

The following table, construed mainly from recent INFCE studies devoted to waste management, provides a rough comparison between conditioned waste quantities produced by successive steps of both LWR and FBR fuel cycles, for a 1 GWe·a electricity production:

Conditioned waste volume m^3	1 GWe·a LWR	1 GWe·a FBR
Uranium mill tailings	58,000	340
Conversion and enrichment plants	73	4
Fuel fabrication plant	40	64
Reactor waste:		
-Operation	400	6
-Maintenance	55	94
-Control rods	1.8	4.2
Reprocessing plants:		
-Nulls, spacers, insolubles	20	52
-Vitrified high-level waste	5	4
-Noble gases	1	1
-Medium-level and plant maintenance waste	28	13
-Low-level waste	130	130

Generally speaking, waste management problems encountered in both FBR and LWR fuel cycles are similar, both for waste characteristics and for waste quantities, at least for steps having an equivalent part in both cycles:
fuel fabrication, reactor operation, fuel reprocessing.

Nevertheless, some specific aspects of FBR fuel cycle waste management can be briefly pointed out:

FBR fuel fabrication wastes are somewhat higher than LWR fuel fabrication ones, and include an important part of α contaminated wastes([241Am], Pu). Moreover it is highly desirable to recover as far as possible the plutonium included in the wastes (presently 1% to 2% of the total Pu amount) which would be otherwise lost for fuel fabrication,

as it is well known, FBR operation produces less wastes than LWR operation,

FBR fuel reprocessing produces more wastes than LWR fuel reprocessing, what concerns hulls, spacers, insolubles. This is mainly due on the one hand to stainless steel cladding and wire spacers, and on the other hand to higher burnup rates reached for FBRs, which gives rise to more insolubles (platinoids among fission products).

7 Conclusion

Out-of-pile steps of the fuel cycle play an essential part in nuclear energy development and their completion at an industrial scale is necessary to nuclear program development.

For what concerns LWRs, fuel cycle activities which are upstream of the reactors: ore extraction and conversion, uranium enrichment, and fuel fabrication, have fully reached their industrial maturity. One has to note that ore extraction, which is performed throughout the world, can provide a diversified and thereby safe raw material supply for countries without significant resources. Nevertheless, in a long-term perspective, one has to remember that uranium resources are limited. In this respect, FBRs appear to be a necessary complement to LWRs insofar as one considers nuclear energy as a long-term means to meet the energy demand. Several industrialized countries have engaged FBR programs aiming at the industrial level by the end of the century.

As to back-end activities of the cycle, i.e., reprocessing and waste management, their development, which is closely related to the strategy choosen for FBR development, is just reaching the industrial level. Large reprocessing capacities ($\approx$1500 t/year) should be operational by the early 90s in industrialized countries (e.g., France, Japan, the United Kingdom, the Federal Republic of Germany).

One has to note that plutonium feeding of the first commercial-size breeders will be produced by LWR fuel reprocessing: approximately 800 tonnes of LWR fuel are necessary to get the first

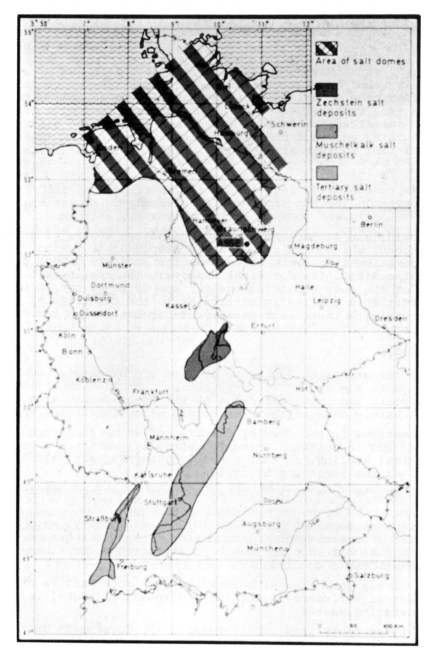

Fig. 1. Permian, Triassic and Tertiary salt formation in the Federal Republic of Germany

the Federal Republic of Germany, in order to test and go ahead
with the disposal of radioactive wastes of all categories within
the frame of extensive research and technical developmental programs.

One of the main issues is the development of suitable and safe disposal technologies. The Asse Mine is therefore a pilot repository. The experiences gained here are important and essential for the planning and construction of the large final repository in the frame of the nuclear recycling and dispoal plant in a North German salt dome.

As site for this recycling and disposal plant the cabinet of Northern Saxony proposed the salt dome Gorleben in the Lüchow Dannenberg area near the river Elbe. On account of seismic investigations the size of the salt dome, namely 38-42 km^2, as well as the thickness of overburden strate of approximately 300 m, are known. In comparison with Gorleben, the Asse salt dome with a size about 4.2 km^2 is relatively small.

Almost completely unknown is, however, the detailed geological structure of the Gorleben salt dome, as the exploratory drillings were only begun in January 1980. To be able to determine the suitability however, an extensive geohydrological investigation is required so as to obtain as exact an idea as possible of the inner structure of the salt dome, as well as of the constitution of the water-carrying overburden strate, in particular in the boundary area Salinar/deeper overburden strata layers. The most important questions in this case are those concerning the structure of the cap rock, the thickness of the rock salt layers which are to be considered solely suitable for the disposal, as well as the question of the existence, position, and thickness of the potash seams, in particular of carnallite (such seams are not suitable for disposal).

Taking the size of the salt dome at Gorleben into consideration it may be fairly definitely assumed that it will be able to fulfill the hopes set on it.

2   Asse Pilot Repository and Planning of Gorleben Disposal Plant

2.1   Geological and Mine Situation

The Asse is a range of hills approximately 8 km long which is situated in the northern foreland of the Harz mountains. It is formed by an asymmetrical anticline of the Triassic age. Below a sedimentary mantle of Buntsandstein and Muschelkalk several hundred meters thick lies the anticline proper, which is formed by strata of the German Zechstein series of the Permian age (Fig.2).

Salt Mine Asse II was initially concerned with the mining of carnallite until 1923 and then only with the exploitation of rock salt. While the potash rooms were backfilled, about 100 mine rooms in the Older and Younger halite, in the range of 15 levels between 490 and 800 m depth, were created by rock salt exploitation. All of these mine rooms represent a volume capacity of approximately 3.5 miilion m^3 (Fig.3).

Nuclear Fusion with Magnetic Containment

A.Schlüter[1]

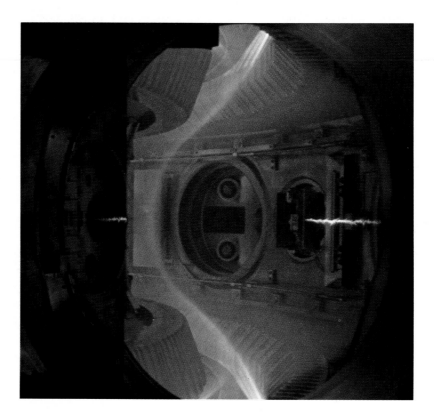

View into the interior of the ASDEX vacuum vessel during a typical tokamak
divertor discharge in natural colour. The bright parts correspond to the
surface part of the plasma which is deflected into the upper and lower divertor chambers which act as "vacuum cleaners". On the right hand side the
trace of an injected pellet consisting of solid deuterium is seen. The left
end of the trace corresponds to the point where this pellet is completely
evaporated. A part of the trace is mirrored on an observational plane mirror

Photograph: ASDEX and pellet-injection team, IPP

1 Max-Planck-Institut für Plasmaphysik, 8046 Garching, FRG
  Lecture held at Kiel University, Jan. 1980

The aim of fusion research is to control the process of nuclear fusion in order to utilize the energy released. The underlying principle is the reverse of that involved in nuclear energy production by nuclear fission. Fusion must be acknowledged as much more natural than fission because the processes aimed at resemble in principle those occurring deep in the interior of all ordinary stars. That is why the author once referred to this objective as trying to do what Prometheus was so severely punished for, viz. taking fire from the sun as a gift for Man on earth.

Now, 25 years after tackling this task, we believe that we are so close to this goal that we could build a first small test reactor if we could find somebody who has the 500 million marks that we have not and shares our opinion that it would be well spent on this objective. Before going into this proposition in detail, let us first return to the principles.

Every form of nuclear energy is sustained by the fact that nucleons, the components of the atomic nucleus, are bound to one another with an intensity depending on the particular nucleus. According to Einstein these different binding energies can be determined directly by weighing. After deducting the weights of the nucleons from the weigths of the various atomic nuclei, it is seen that energy can be gained in principle by fusion of small atomic nuclei to form larger ones and by fission of the largest atomic nuclei to yield medium-sized ones. In attempting practical utilization it is of decisive importance to draw an essential distinction between fusion and fission. In fission the reaction is initiated by penetration of an uncharged particle, a neutron, into the nucleus, a chain reaction being made possible by the fact that further uncharged neutrons are produced by the resulting nuclear fission. In fusion, by contrast, two different nuclei, which are positively charged like all atomic nuclei, have to be brought so close to one another against the repulsive force exerted between like charges that the nuclear forces become effective and the nuclei react with one another. Even nuclei with the lowest charge, viz. the various isotopes of hydrogen, require an energy of between 10,000 and 20,000 eV to overcome the repulsive force. In cases of the most favorable reaction, as presented in Fig.1, the corresponding energy gain involved in the nuclear reaction is about 14 million eV. One might now think that it would be sufficient to build a simple accelerator for accelerating suitable atomic nuclei in an electric field equivalent to a voltage of 10,000 to 20,000 V, and to fire these projectiles at a target composed of the atomic nuclei of the reaction partner. The basic difficulty is that the yield from this process would be so small that the energy gain from the reactions could not exceed the energy input because even in the most favorable reactions and with the closest possible encounters of the atomic nuclei the fraction of collisions leading to reactions would be very small, and in the other cases, i.e., in almost all cases, the atomic nuclei would only deflect one another and therefore the input energy would have to be regarded as lost. There is only one possibility where these ordinary "elastic" collisions do not have to be classed as loss processes, and that is when the kinetic energy required does not occur in an ordered

184

$$D + D \longrightarrow n + {}^3He \quad + 3.267 \ MeV$$
$$D + D \longrightarrow p + T \quad + 4.032 \ MeV$$
$$D + T \longrightarrow n + {}^4He \quad + 17.58 \ MeV$$
$$D + {}^3He \longrightarrow p + {}^4He \quad + 18.351 \ MeV$$
$$T + T \longrightarrow 2n + {}^4He \quad + 11.328 \ MeV$$

Fig. 1. Fusion reactions. The reaction most likely employed is the one of
the *3rd line*, although tritium, the superheavy isotope of hydrogen, does
not occur in nature and must therefore be bred in the blanket of the fusion
reactor. The reaction of the 4th line might also be technically feasible,
however, the isotope [${}^3He$] of helium is also extremely rare in nature and
no way of producing it is known. The reactions of the *first two lines* would
not only provide energy but also fuel for the mentioned reactions. Their
disadvantage lies in the small reaction probabilities which seem to exclude
the technical feasibility

particle beam but in unordered motion corresponding to a sufficiently high temperature. The temperature needed can easily be
calculated. It is about 100 million degrees, which is really
high. To give some idea of how high a temperature of 100 million
degrees is, it is again convenient to enlist the sun as an il-
lustration. If the temperature at the surface of the sun were
not about 5000 degrees but 100 million degrees, it would only
need a diameter of 3.5 m instead of its 1.5 million km to radiate
the present quantity of energy to earth. Moreover, a ball of this
size would be very roughly equivalent to the volume which we
would in fact have to raise to this temperature in a future fusion reactor.

This leads us straight to another principle. The hot gas needed,
plasma as it is called, ought not to radiate at anywhere near
the intensity corresponding to its temperature. According to
the known physical laws this means that it has to be very highly transparent because the less a body absorbs the less it
radiates. One of the prime questions is thus whether the absorption is in fact low enough. In astrophysics this is a standard calculation. It is found that in the case of hydrogen,
which at these high temperatures completely disintegrates into
its singly charged nucleus and one electron, the absorption,
and hence the radiation, is in fact low enough, but the margin
is not very big and even small quantities of foreign substances
such as oxygen or metal impurities would enhance the radiation
to such an extent that the energy produced by the nuclear reaction could not maintain the burn temperature.

Radiative energy loss is not the only problem, however. Ordinary
thermal conductivity between any material wall and the plasma
would result in such a high energy flux that the surface of the
wall would immediately be sputtered, thus cooling down the plasma. There is only *one* way of keeping the plasma away from a material wall for times longer than, say, 1 milliardth of a second,
and that is to apply magnetic fields. The same property which
makes thermonuclear fusion so difficult in the first place, viz.
the electric charge of the reaction partners, is exploited here
because it also produces both active and passive interaction

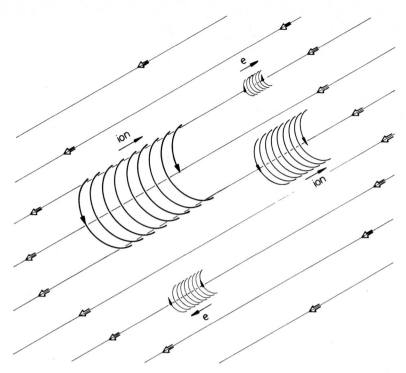

<u>Fig. 2.</u> Typical trajectories of charged particles in a uniform magnetic field

with a magnetic field. As indicated in Fig.2, charged particles
are bound to spiral paths around magnetic field lines, moving
freely along them. If, therefore, the plasma is placed in a magnetic field so that no field line with plasma located anywhere
on it leaves the plasma and only unoccupied magnetic field lines
intervene between the plasma and material walls, it is possible
to achieve the desired insulation, and it does not require much
thought to see that there is hardly any other way of doing this
than to make the plasma ring-shaped, i.e., to form a plasma torus.

This is where new difficulties start because the tying of the
electric particles to the magnetic field lines is strictly valid
only if the field is uniform and its lines are straight, and of
course they have to be bent to form a close, ring-shaped or
"toroidal" field. The particles can, however, be successfully
confined despite the ring curvature if the curved field lines
are also twisted so that they all entwine around a circularly
closed line, which is referred to as the magnetic axis.

Let us now take a quick look at theory. The twisting of the field
lines can be achieved in two ways — the alternatives having been
recognized independently in the United States, in the U.S.S.R.,
and by us about 25 years ago — which have led to two families
of experiments, the tokamak and the stellarator, as they are
now generally called. In the tokamak principle the field lines
are twisted by utilizing the "active" component of the interac-

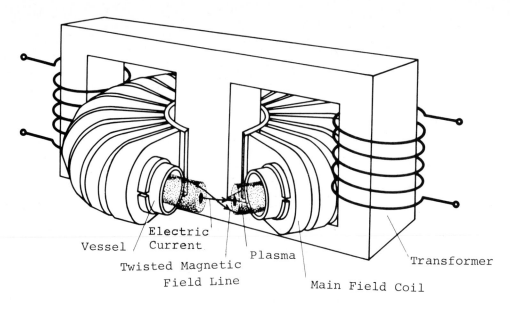

Vessel / Electric Current

Twisted Magnetic Field Line

Plasma

Main Field Coil

Transformer

Fig. 3. Tokamak principle. Decisive for the confinement of the plasma is the electric current induced in the plasma by a transformer

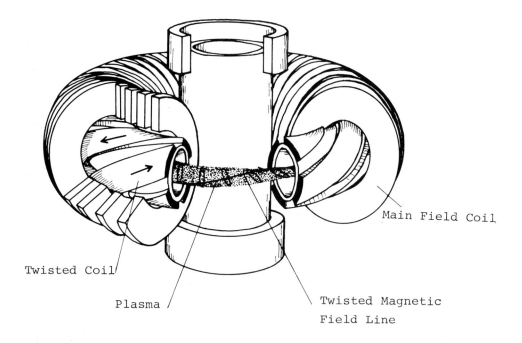

Main Field Coil

Twisted Coil

Plasma

Twisted Magnetic Field Line

Fig. 4. Stellarator principle. Decisive for the confinement of the plasma is the twist of the magnetic field lines which is produced by special twisted coils

action between charged particles and magnetic fields: ordered
motion of charged particles constitutes an electric current
which generates a magnetic field. If, in particular, the current
is made to flow along the ring, i.e., in the torroidal direction,
the field of this current produces the desired twisting of the
externally generated main field. Figure 3 shows the coils that
generate the main field as well as the iron yoke of the transformer and its primary windings. The plasma ring forms the secondary circuit. In the stellarator, by contrast, complicated
coils are needed to achieve twisting externally as well (Fig.4).
One presumed advantage of the stellarator is that it affords
better stability; this is important since magnetic confinement
is always a very "delicate" matter. The particles of the plasma
keep finding ways of moving collectively to change the magnetic
field and thus escape from confinement. It is not our intention
here to give a detailed comparative assessment of these two
principles (nor even to mention other proposals). Let it just
be stated that since the very successful work performed on tokamaks at the Kurchatov Institute about 15 years ago good progress
has been made throughout the world with increasingly large
versions of this concept; recently, however, stellarators have
been catching up and may yet finally represent the better fusion
reactor.

At Max-Planck-Institut für Plasmaphysik (IPP) in Garching, on the
outskirts of Munich, we are fairly neutral. With ASDEX (Fig.5)
we have had the largest tokamak in Europe outside the Soviet
Union in operation for a year. This tokamak has yielded interesting results. In fact, it holds a few world records, though this
is not what it was built for. Its purpose is to prevent impurities from entering the plasma, since even small quantities will
contaminate it — owing to the radiation properties already mentioned. There appear to be ways — to which ASDEX can contribute —
of solving this important and difficult problem in a "reactor
relevant" manner.

We also have in operation one of the few large stellarators,
ours being the only one which (for almost a year) has really succeeded in confining a relevant plasma (i.e., hot and dense)
"from outside". Confinement here really does seem to be better
than in similar tokamaks (considering the small value of its
cross section). At present we are calculating new field configurations which may further improve the behavior of the plasma.

As already stated, the stellarator still has to catch up on the
tokamak. This implies that there is some measure by which to
determine the distance to the objective. This can in fact be
done with figures, though at the price of gross simplifications.
Among the figures characterizing the status quo on the way to
the fusion reactor two are obviously of particular importance,
describing the achieved temperature and the quality of the confinement respectively. For any state of a confined plasma the
latter compares the energy input rate necessary to maintain the
temperature with the energy content of the plasma. The energy
containment time τ thus defined needs to be the larger, the
smaller the gas pressure p of the plasma. So the product $p \times \tau$

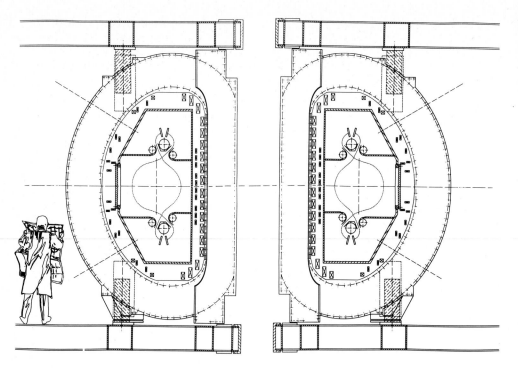

Fig. 5. Cross section through the ASDEX device. The plasma ring is limited by the *broken line* which, after passing through a "bottleneck", end on collector plates. This relatively complicated scheme shall improve the isolation of the plasma and reduce the flux of impurities into the plasma

is a fair figure of merit. In Fig.6 this product and the temperature are shown for a few experiments and are compared with the minimum confinement quality for a fusion reactor, in which it is required that the energy of the electrically charged products (α particles) of the fusion reactions be used — and be sufficient — to maintain the burn temperature.

Particular mention should be made of the values from PLT (Princeton Large Tokamak) and Alcator (Alto Campo Toro, Massachusetts). In the case of ASDEX allowance has to be made for the fact that the main heating system is still being built. Figure 6 also gives the values predicted for two devices that do not yet exist. The first is the JET (Joint European Torus) experiment now being built by the European Community at Culham in England. The so-called Extended Performance Version is at present being prepared and is to be put into operation in the middle of 1987. It is this version to which the data refer. A year later JET is to be filled with the deuterium-tritium mixture, and will then probably produce a large number of fusion reactions — perhaps these will be sufficient to maintain the temperature for a short time. It would then be not only the first machine to work as an "ignited" fusion reactor but also, as far as present planning goes, the only one for the next few years. The second device represented is ZEPHYR, a machine designed at IPP, which is discussed below.

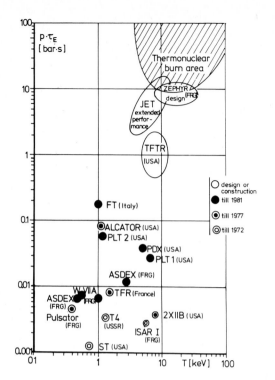

Fig. 6. Temperature and quality of confinement. The necessary values for self-sustaining D-T thermonuclear reaction are indicated by the *hatched area* and compared with results achieved in different machines and expected for the tokamak TFTR under construction, the extended performance version of JET in preparation, and of the ZEPHYR design. The diagram is updated to represent the state of end 1981. *Note*: 1 bar ≈ 1 atmosphere; 1 keV ≈ 10 million degrees C

As far as these two figures of merit indicate we are thus on the right road toward this goal. Looking back at the progress achieved over the years, one is even tempted to predict already when the final success will come. These figures really give cause for optimism, but, as already stated, they should not be allowed to obscure the physical and technical problems involved. One physical problem, for example, is that the heat transport from the inside to the outside in a magnetically confined plasma is actually greater than predicted by the theory and is based on processes not yet understood. It is nevertheless still small, and presumably small enough to make a fusion reactor possible. This discrepancy, however, introduces an element of uncertainty into the prediction which not only affects operation of the fusion reactor but also has a direct bearing on the design of the heating for the next set of experiments because the higher the heat transport, the higher of course the heating power has to be; and it is really high powers that are involved in the first place. A figure should be mentioned: The power required to heat the plasma of the JET experiment to the thermonuclear regime is estimated at 25 MW, and this has to be deposited in the core of the plasma. This entails a large complex of technical problems. The theory of plasma heating, e.g., by confinement of high-energy particles or by high-frequency waves, has not yet been fully developed either and the available experimental data are incomplete, but promising on the whole. Obviously, however, the actual behavior of the plasma essentially depends on the heating method. A burning fusion reactor plasma, which by definition is heated by the alpha particles produced, must therefore be expected to behave differently to plasmas heated in other ways, these being the only kind which could be investigated hitherto. The ZEPHYR experiment planned at IPP is intended to fill this important gap. With favorable assumptions on the heat transport, particle confinement, and effect of the heating methods envisaged, the plasma produced in ZEPHYR would really burn in the thermonuclear regime, i.e., maintain or even increase its temperature by itself after the external heating is switched off, as long as the extremely high magnetic field in ZEPHYR can be maintained or till the burn process is extinguished by hitherto unknown dynamic processes or simply the accumulation of impurities. The purpose of ZEPHYR is precisely to study the dynamics of the burn process and the behavior of impurities. These investigations would have to be done on the actually burning plasma before tackling the construction of a fusion reactor close in size to future power reactors that will cost several thousand million dollars. The impurity problem here is so complex and detrimental that it already has to be given very close attention now, this being the main concern of ASDEX, the large tokamak mentioned above.

Research hitherto has been mainly concerned with questions of plasma physics because it is here that the biggest step into the unknown has to be taken in building a fusion reactor. It is also here that one is most likely to discover that the necessary temperatures or degrees of purity cannot be achieved, i.e., that the physics do not work. It has already been indicated above that experimental progress increasingly confirms the likelihood that the fusion reactor will not be condemned to failure despite these difficulties. It then remains to solve the technical problems and, finally, to demonstrate the superiority of the fusion system with respect to its economy and its environmental impact so that fusion can really assume a major role in solving the energy problem.

The technical problems here are immense. The most serious of these are due to the neutrons produced by the fusion processes, because the neutron nuclei react with a large number of materials. It is the neutrons that contain most of the fusion energy, and so it is therefore primarily their energy that has to be economically utilized. They are relatively much more numerous and also have a higher energy, thus being more harmful, than those in fission reactions. Neutrons are also indispensable for breeding, i.e., producing the fusion fuel, tritium, from lithium metal. This entails an as yet unsolved materials problem, because the maximum wall load that can be withstood by materials whose properties under neutron bombardment are known is at most the equivalent of a primary-neutron energy flux of about 2 MW/m^2. Even at this low level the first wall, which is also exposed to all the other kinds of radiation from the plasma, can only be expected to last for a few years and will then have to be replaced. The power density resulting from this relatively low energy flux is also much smaller than that in fission reactors, particularly that of the fast breeder. This means that for a given power plant output the actual nuclear section in a fusion

reactor will be considerably larger than that in a fission reactor. This need not necessarily rule out the economicalness of
the fusion reactor, since the lower power density ought to make
it easier to remove heat from the nuclear section, i.e., to cool
the so-called blanket. The need to replace the first wall incurs
further design complications since it is of course located inside the magnetic field, which is generated by refrigerated super-
conducting coils, and somehow these have to be pushed aside to
afford access to the first wall. All this has to be accomplished
in a highly radioactive environment, i.e., by remote handling.
Our investigations for the ZEPHYR project, in which these questions already arise, have demonstrated that a technique for per-
forming such operations reliably and quickly by remote handling
still has to be devised. It would then, moreover, also be of
great use for repairs and maintenance in conventional nuclear
power plants.

Finally, the neutrons can also be utilized to convert the non-
fissile isotopes of uranium and thorium (which, unlike fissile
isotopes, are abundantly available) into fissile substances
which can be used in conventional fission power plants. No experiments have yet been conducted on fusion-fission hybrid systems, but calculations indicate that a fusion reactor may probably be capable of supplying about ten fission reactors, each
having a thermal power equal to its own. This affords advantages
over the use of fast breeders, e.g., it is then no longer necessary to provide each breeder with its own fast fission fuel load.

What then are the prospects of fusion research? It seems that
the physical problem of fetching fire from the sun to the earth
can be solved. But even with favorable forecasting it will take
another generation of large-scale experiments beyond the JET or
ZEPHYR type before construction of a demonstration reactor can
be considered, which means that final proof of feasibility can
scarcely be provided before the turn of the century. At the same
time we shall need a technology program calling for greater out-
lay than the effort expended in solving problems of plasma physics. Economically acceptable solutions to the fusion problem
will therefore definitely not be available till well after the
turn of the century.

Why then all the effort and expense? It is because nuclear fusion
represents a worthwhile option for solving the energy problem,
both in generating electricity and in producing useful heat,
for which the primary fuels, viz. heavy hydrogen or deuterium,
which is abundantly present in any kind of water, and lithium
metal, are generally accessible. The fertile material required
for the hybrid reactor is also readily available. In view of the
worldwide energy problem, which will certainly get worse, this
option can hardly be neglected. Furthermore, fusion energy and
fission energy both have the property of being less harmful to
the environment than, say, forms of energy using fossil fuels.
The greatest environmental problem, as in fission power plants
(with the possible exception of high-temperature reactors),
will probably be the thermal impact on the environment when the
nuclear heat is converted into electricity. The safety problem
seems less difficult then in fission reactors, thus allowing a

that of a comparable fission reactor. (4) The radioactive tritium
is processed inside the reactor in a closed cycle. This inhibits
or even prevents misuse of the radioactive material.

It is therefore worthwhile making any effort to develop this new
energy source. This is especially important for the European
countries to make them more independent of oil and other fuel
imports.

However, it is not simple to realize a fusion reactor. The nuclei of the fuel have to overcome repulsive electrostatic forces
and get so close to one another that nuclear interaction forces
prevail and fusion can occur. This can be done hy heating a mixture of deuterium and tritium to temperatures of 100 million
degrees. By keeping such a hot "plasma" together sufficiently
long, enough fusion collisions are produced and more energy is
set free by fusion than was supplied before to heat and ignite
the fuel.

Walls of solid materials cannot stand these high temperatures.
However, such a hot plasma consisting of charged particles, ions
and electrons can be confined by suitable magnetic fields. Investigations to heat and confine fusion plasmas in magnetic
fields started in the mid-1950's. Encouraging progress has been
made especially in the last few years in the field of TOKOMAK
configurations. Details on this were given in a previous report
of this series.

In spite of this progress, however, it is not sure that a fusion
reactor on the TOKOMAK principle producing power economically
will be possible. Other novel paths to a fusion reactor are
therefore still welcome. This is true especially of new ideas
that avoid the complicated and expensive magnetic field coils.
The advent of the laser has opened up such a new path toward
a fusion reactor. Modern high-power lasers can produce radiation
powers of several TW (1 TW = 10^{12}W) in a single beam in short
pulses with a length of about 1 ns or less (1 ns = 10^{-9}s). If
this radiation is concentrated with a lens, one gets power densities above 10^{16}W/cm^2 in the focal region. This corresponds to
an electric field strength of the light wave of more than 10^9V/cm,
which immediately ionizes any matter and transforms it into a
hot plasma. Focusing of the laser beam in air or another gas
produces a so-called laser spark (Fig.1). Temperatures in the
range of 10^5 K can easily be achieved in this way, but this is
not high enough to initiate fusion reaction. If the power of
the laser is increased, so is the volume of the spark but not
the temperature. In order to get the ignition temperature of
10^8 K for fusion reactions, the laser energy has to be put into
a limited amount of matter. This can be done by using a small
pellet of solid hydrogen fuel and irradiating it with a short
but energetic laser pulse. The principle of such laser-ignited
fusion is shown in Fig.2: A solid (frozen) DT pellet is injected
into a reactor chamber. When this pellet passes the center of
the chamber a number of powerful lasers are triggered. Their
energy is concentrated onto the pellet, transforms it in to a
hot and dense plasma and ignites fusion reactions. At the same
time the pellet starts to expand. Because of the high density,

Fig. 1. A laser beam coming from the right side is focused by a lens and produces a laser spark in the focus

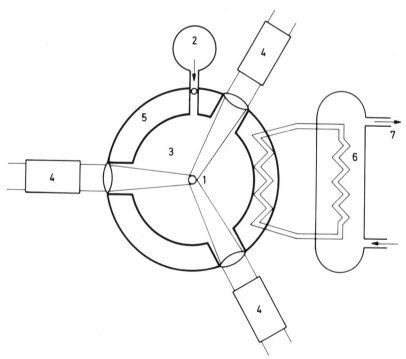

Fig. 2. Principle of a laser fusion reactor. *1* DT pellet; *2* pellet injector; *3* reactor chamber; *4* lasers; *5* reactor blanket with lithium; *6* boiler; *7* steam to turbine

many fusion reactions occur during the short lifetime of the plasma and an excess of energy is produced. As it is only the forces of inertia which tend to keep the plasma together, the term "inertial confinement" is also used instead of laser fusion. The energy is collected on the wall and in the blanket of the reactor cavity. The blanket contains lithium or a lithium compound in which the fusion neutrons are stopped, transforming their energy into heat or reacting with the lithium according to Eqs. [2] and thus producing the tritium fuel. Heat exchangers are used to produce steam for driving a turbine.

The principle of the laser fusion reactor is attractive because of its simple geometry and because complicated magnetic field coils are not necessary. However, there are a number of questions which have to be answered before a definite prediction on the feasibility of an economically operation laser fusion power plant can be made.

2 First Question: How Much Energy is Required to Ignite the Pellet and What is the Energy Gain?

Clearly, the basic requirement is that the fusion energy gained at each "shot" must exceed the laser energy used to ignite the pellet. Two conditions have to be attained in order to reach ignition:

$$\text{plasma temperature} \approx 10^8 \text{K} \quad , \tag{3}$$

$$\rho \cdot R \geq 0.3 \text{ g/cm}^2 \quad . \tag{4}$$

The first condition has already been introduced above. The second condition (ρ is the mass density of the pellet plasma, and R its radius) is to ensure that the range of the α-particles produced by the fusion reactions is smaller than the pellet plasma radius, i.e., the α-particles from the reactions are stopped inside the fuel and provide additional heating. If it is assumed that the pellet is heated in a very short time, then the initial density ρ of the plasma produced is equal to the density of the solid fuel, i.e., approximately 0.2 g/cm^3. To meet condition (4), the radius of the pellet must be of the order of 1 cm. The corresponding heating energy Q_h to be supplied by the laser is

$$Q_h \sim \rho R^3 \tag{5}$$

and can be calculated for this case to be of the order of 10^6 kJ. This amount of energy is much higher than even optimists could hope to achieve with future lasers. In addition, the amount of fusion energy delivered from such a large pellet would make it impracticable for a reactor. However, inserting the limiting case of (4)

$$\rho R = 0.3 \text{ g/cm}^2$$

into the relation (5) yields

$$Q_h \sim \rho \frac{1}{\rho^3} = \frac{1}{\rho^2} \quad . \tag{6}$$

This means that the heating energy can be essentially decreased by a higher density of the fuel, i.e., the fuel has to be compressed to multiples of the solid-state density. But energy also has to be invested for compression, and the ignition energy Q is now the sum of the compression energy Q_c and heating energy Q_h:

$$Q = Q_c + Q_h \quad . \tag{7}$$

As the compression energy Q_c increases with increasing density, whereas the heating energy Q_h decreases, there is an optimum compression ratio for the achievement of the smallest ignition energy. Various extensive numerical calculations indicate that this optimum ranges between 10^3 to 10^4 times the solid-state density. Early estimates have shown that ignition energies as low as 1 kJ should be sufficient to reach "break even", i.e., to get as much fusion energy out of a pellet as was used before for its ignition. However, in the meantime it has become clear that the efficiency of the compression process is a major factor which has to be taken into account. Therefore, in order to give a more realistic answer to the first question, one has to consider the compression process in more detail.

3 Second Question: How Can the Fuel Be Compressed and What Is the Efficiency of the Compression Process?

So far we have talked about "heating of a pellet by laser pulses" in very simplified terms, implicitly suggesting that the pellet is homogeneously heated by the laser radiation. Closer scrutiny reveals, however, that fast heating of solids by lasers is a very complex process. Specifically, it is always connected with hydrodynamics and consequently with pressure and density changes. But it turns out that this very complexity favors laser fusion.

In order to understand the essentials, we consider a spherical pellet of solid hydrogen (Fig.3). This pellet is homogeneously irradiated from all sides by a sufficient number of strong lasers. In a very short time after the lasers are switched on, a thin layer at the pellet surface has become fully ionized and has an electron density larger than the critical density[2]. The laser light cannot penetrate into the center of the pellet and is absorbed (and partially reflected) in the outer layer. This layer is heated to a high temperature, a high pressure is built up and causes radial expansion. By this process the density of the layer decreases below the critical value, and the light can

---

2 The critical density of a plasma is defined as the density where the electron plasma frequency is equal to the light frequency. Electromagnetic waves can only propagate in plasmas with densities below the critical density (more details below)

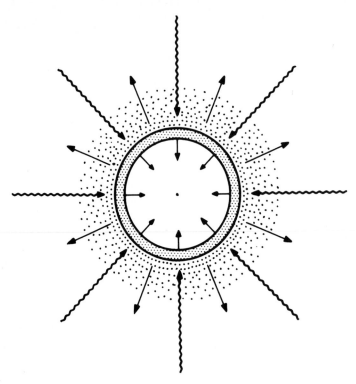

penetrate more deeply into the pellet and heats the next layer. In this way the pellet is continuously evaporated layer by layer in what is called an ablation process. This seems to be adverse to our aim, viz. compression and heating of the fuel. However, the ablation process has a beneficial side, too: The reaction forces of the radially (outwardly) expanding plasma cause a compression wave traveling radially inward toward the center of the pellet. By proper programing of the laser intensity in time, nearly isentropic compression of the remaining core of the pellet can be achieved. The temperature in the core increases correspondingly and ignition temperatures can be reached in the center. In this ablatively driven compression process a large portion of the pellet mass is evaporated. This part therefore need not be solid DT fuel but could be made of other materials which might be more favorable with respect to light absorption or hydrodynamics. In fact, pellets consisting of one or more spherical layers of different suitable materials seem to promise a much higher energy gain (i.e., the ratio between fusion and ignition energy) than a simple solid DT pellet. In addition, the requirements imposed on the behavior of the laser pulse can be essentially relaxed by properly designed pellets.

The processes qualitatively described above are extensively investigated by means of large computer codes in which many of the relevant physical phenomena are included. On the basis of such computer calculations from different laboratories we can now answer our first question: The laser energy necessary for the ignition of pellets in a fusion reactor is of the order of 1 MJ. This energy has to be delivered in pulses with a length between

1 and 10 ns, i.e., with peak powers of 10^{15}W. The energy gain
will then be between 100 and 1000, or 10^8 to 10^9 J of fusion
energy will be produced from a single pellet with a typical diameter of a few millimeters. If a reasonable repetition rate of
10 pellets per second is assumed, such a laser fusion reactor
would have a thermal power of 1 to 10 GW, this being comparable
with power stations now in operation.

4 Third Question: What Type of Laser Is Suitable for Driving a Fusion Reactor?

We have seen that a light energy of about 1 MJ in pulses of 1 to
10 ns has to be delivered to the fuel for ignition. In order to
get symmetric compression, the pellet has to be homogeneously
illuminated from all sides. At least two separate laser beams
irradiating from opposite sides therefore have to be used, but
probably a greater number may be necessary for technical reasons. One of the main points of concern of the reactor designers
is the efficiency of the laser, i.e., the ratio between the energy of the light pulse and the electric energy input into the laser.
If a fusion energy gain between 100 and 1000 in the pellet is assumed, a laser efficiency of greater than 1% up to 10% is required for a net energy output. Design studies have further
shown that for a power plant the laser has to be fired one to
ten times per second for reasons of economy. As transfer of radiation energy into compression and heating of the fuel is more
efficient at shorter wavelengths, laser should work at wavelengths
as short as possible. A laser that meets all the requirements
for an economically working power plant does not exist at present.
However, there are a number of laser systems which are now available, and which can be used for experimental study of the physics
of laser fusion. Before describing the three main types of fusion
lasers, Nd-glass, CO_2, and iodine, and their present technical
status, let us have a short look at how a laser functions in
principle.

5 Question: How Does a Laser Work?

Let us remember the structure of an atom in its simplest form:
It consists of a heavy nucleus with positive electric charge.
Electrons travel around this nucleus on discrete orbits characterized by certain energy values. Light is emitted if an electron jumps from one orbit to another orbit of lower energy. The
frequency ν of the light (its colour) is determined by the energy difference ΔE between the two orbits: $h \cdot \nu = \Delta E$ (h = Planck
constant). The frequency ν of the emitted light increases (its
wavelength decreases) with increasing magnitude of the energy
jump.

In order to understand how a laser works, we consider a very
simplified model of an atom with only two orbits for one electron as illustrated in a two-level energy diagram (Fig.4). The
ground state corresponds to an orbit with the lowest possible
energy, the excited state to an orbit with a higher energy. At

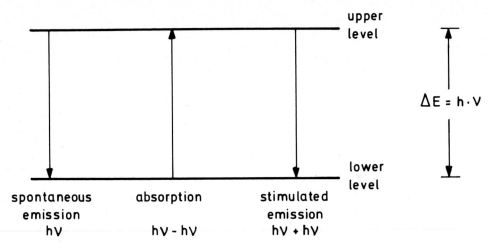

Simplified two-level energy diagram of an atom. Transition from
the upper to the lower level is connected with radiation, transition from
the lower to the upper level with absorption of light frequency

room temperature in a gas volume containing a great number of
our model atoms, practically all the atoms are in the ground
state. When the gas is heated to a higher temperature, some of
the atoms are transferred to the excited state by collisions.
Their lifetime in this state, however, is limited. After a certain time which is characteristic of the specific transition
the electron "falls" back into the ground state "by itself" or
spontaneously. Light is thereby emitted with a frequency ν corresponding to the energy difference ΔE of the two levels. The
gas radiates. This process is called spontaneous emission.

The transition from one energy state to another can, however,
also be caused by irradiation of light: An atom in the ground
state absorbs light of frequency $\nu = \Delta E/h$ and is thereby transferred into the excited state. If an atom happens to be in an
excited state, the incoming light stimulates an "earlier" transition to the ground state and additional light is emitted.
Thus, when light of frequency ν is transmitted through our gas,
it is partially absorbed, its intensity being reduced if most
of the atoms are in the ground state (=normal condition). If,
however, the population is inverted", i.e., more atoms are in
the excited state than in the ground state, then the original
intensity of the light is increased by stimulated (or induced)
emission. In the latter case we have got a light amplifier, a
LASER (Light Amplification by Stimulated Emission of Radiation).
Generally speaking, in order to realize a laser, one has to
produce "inversion" in a proper medium, i.e., an abnormal condition in which more atoms are in the excited than in the ground
state.

This light amplification can be used to produce light intensities never known before. For this purpose one constructs a multi-
stage system consisting of an oscillator and a series of consecutive amplifiers (Fig.5) (analogous to a radio-frequency emitter)

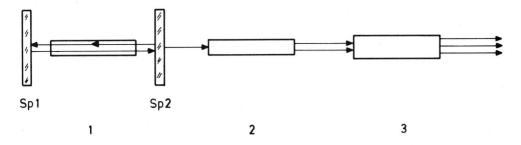

Sp1   Sp2

1   2   3

Fig. 5. A light pulse is produced in an oscillator *1* (*Sp 1*, *Sp 2* are partially reflecting mirrors) and is then amplified in the consecutive amplifiers *2* and *3*)

The oscillator is essentially built from a tube containing the
laser medium and two mirrors. Inversion is produced by a proper
pumping process; some spontaneous emission occurs isotropically
in all directions. A small portion of this spontaneously emitted
light is perpendicularly incident on the mirrors and reflected
back and forth. On every path through the inverted medium it is
amplified. In this way, a strong light pulse is built up from
the weak spontaneous emission. Part of the light pulse is transmitted through one of the mirrors which is partially transparent
and is further amplified on its way through a series of tubes
containing the inverted laser medium, the amplifiers.

The final laser pulse has a number of specific properties that
discriminate it from "natural light." Lasers are therefore of
interest for a wide variety of applications. For our purpose,
i.e., laser fusion, only three of the outstanding features of
this new tool are used: the high radiation power delivered in
a highly collimated beam and the short pulse length.

The technically most developed laser system is the Nd-glass
laser. The active medium is ions of the rare earth element neodymium embedded in glass. Inversion is provided by "optical
pumping": A Nd-glass rod (Fig.6a) is surrounded by a helical
flashlamp. Flashlight in the visible part of the spectrum is
absorbed and produces — via some intermediate steps — overpo-
pulation of the upper laser level, thus leading to induced
emission in the near infrared region at 1.06 μm. The diameter
of the laser beam has to be increased as it passes through the
amplifiers in order to avoid damage as a consequence of the
high power densities. As larger diameter rods can no longer be
homogeneously pumped, Nd-glass discs are used for the final amplifiers pumped by linear flashlamps as shown in Fig.6b. Large
laser systems of this type have been developed at several places
in the world. The largest Nd-glass system is in operation at
Lawrence Livermore Laboratory in California (Fig.7). It consists of 20 beams, each with an aperture of 20 cm, delivering
a total light power of up to 30 TW in pulses of 10^{-10} s or up
to 20 kJ in pulses of 10^{-9} s. An installation ten times more
powerful called SHIVA NOVA is under construction, being designed
for all final energy of 300 kJ, only one order of magnitude below the energy level now thought necessary for a reactor. Unfor-

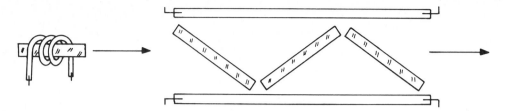

Fig. 6. Nd laser pumping configurations. *Left*: Nd-glass rod surrounded by a helical flashlamp realized for rod diameters up to 12 cm. *Right:* Elliptical Nd-glass discs, for larger beam diameters pumped by linear flash lamps

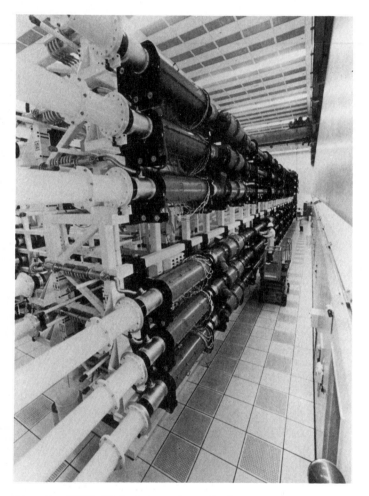

Fig. 7. SHIVA Nd laser system of the Lawrence Livermore Laboratory. View of 6 of the 20 beam lines with the 20 cm aperture final amplifiers in the foreground. (Courtesy of Lawrence Livermore Laboratory)

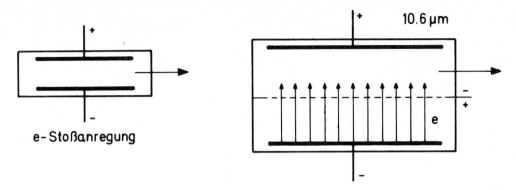

Fig. 8. The CO_2 laser is pumped by electron collisions realized in a transverse gas discharge for smaller cross section (*left*) and an electron beam sustained discharge for beam diameters up to 50 cm (*right*)

tunately, however, Nd-glass lasers will not be suitable for driving a reactor for essentially two reasons. First, the efficiency of this laser is below 0.1% and it is doubtful that 1% or even more can be reached in the future. Second — and this is a major obstacle — the repetition rates of this laser type are too small for a power-producing reactor. The losses not to be avoided in the pumping process lead to heating of the laser glass and this causes optical distortions of the beam. The cooling time is essentially determined by the heat conductivity of the glass. As a consequence, one has to wait about 1/2 to 2 h between two shots in the present SHIVA installation to allow cooling and temperature stabilization.

The limitations of efficiency and repetition rate are much less stringent for the CO_2 laser. Its present efficiency is of the order of 1% and there is potential for further increase. Its active medium is a gas and this can be circulated through an external system for providing efficient cooling. Extrapolations to repetition rates required for a reactor therefore seem reasonable. Its main drawback is though to be its long wavelength of 10.6 µm, which reduces the efficiency of the coupling of the light energy into useful plasma energy for compression and heating.

The active medium of the CO_2 laser is a mixture of helium, nitrogen, and CO_2 gas. The inversion is achieved by electronic excitation (Fig.8). In the case of the oscillator and the first amplifiers a conventional gas discharge between long electrodes is used. In the large-cross-section power amplifier stages the laser gas is homogeneously ionized by energetic electrons. These electrons enter the laser chamber through a thin metal foil on a stable grid that forms the anode of the electron gun and also the cathode for a non-self-sustained discharge in the laser gas. The discharge voltage is chosen such that the electrons in the gas volume are accelerated to an energy optimal for the excitation of N_2 molecules. This excitation energy is then transferred to the upper laser level of the CO_2 molecule and creates inversion.

Fig. 9. HELIOS 8-beam CO_2 laser in operation at Los Alamos Scientific Laboratory. The final amplifiers shown in the picture are arranged in dual modules. Eight sets of mirrors concentrate the beams on the pellet in the target chamber (*center*) (Courtesy of Los Alamos Scientific Laboratory)

CO_2 lasers for fusion experiments are being developed mainly at the Los Alamos Scientific Laboratory in New Mexico. Figure 9 shows a model of the HELIOS 8-beams CO_2 laser now in operation at Los Alamos. Each beam has a diameter of about 34 cm and delivers an energy of 1.25 kJ in pulses of 10^{-9} s duration. In this way up to 10 kJ of light energy at 10.6 µm can be concentrated on pellets situated in the target chamber visible in the center of the picture. The next step in the development of CO_2 lasers is the ANTARES 100 kJ system now under construction and scheduled for completion in 1984.

A younger competitor to the Nd-glass laser is the iodine laser. It has the advantages of being a gas laser like CO_2, but has a more favorable wavelength of 1.3 µm, this being comparable with that of the Nd-glass laser. The active medium of the iodine laser is an iodide such as C_3F_7J. This molecule is dissociated by flashlamps into a radical C_3F_7 and an iodine atom J in an excited state. As there are no iodine atoms in the ground state, inversion is produced by this photodissociation pumping process, and light amplification by stimulated emission is possible.

An amplifier unit essentially consists of a quartz tube containing the active medium and of a number of linear flashlamps arranged around the tube parallel to its axis. In order to control the amplification of an amplifier, foreign gases such as argon are added to the active medium[3].

---

3 The amplification of an amplifier is determined by the product of the inversion density and the stimulated emission cross section σ. The latter can be influenced by pressure broadening of the laser transition. By adding a foreign gas of proper pressure, the value of σ, and consequently the amplification, can be adjusted to an optimum number without changing the amount of inversion energy stored in the amplifier

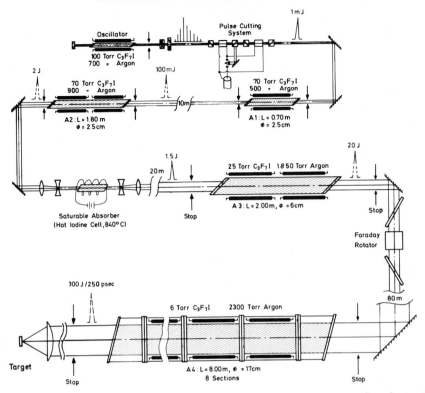

SET UP OF THE IODINE LASER ASTERIX III

Fig. 10. Scheme of the ASTERIX III iodine laser at Max-Planck-Institut für Quantenoptik in Garching. A train of ns pulses is produced in an oscillator (*upper left*). A pulse cutting system selects one single pulse which is further amplified in four consecutive amplifiers

The scheme of the ASTERIX III iodine laser developed at the Max-Planck-Institut für Quantenoptik in Garching near Munich is shown in Fig.10: In an oscillator a series of pulses is produced by "mode-locking" technique. A single pulse is selected and transmitted through an electrooptical pulse cutting system and further amplified in four consecutive amplifiers. At the exit of the last amplifier (Fig.11) the beam diameter is about 16 cm and the maximum power is 10^{12} W delivered in pulses of 0.3 ns duration. Every 10 min such a shot can be fired onto a target where absorption and energy transport processes in the hot plasma are studied. The repetition rate can be increased by implementing a more powerful charging system for the capacitor banks feeding the flashlamps and a larger circulation system for cooling, partially replacing and purifying the laser gas. Repetition rates of up to 10 Hz seem to be technically feasible. The beam quality, i.e., essentially its focusability is not affected by the non-linear index of refraction, which is a serious problem in solid-state lasers. The divergence of the beam is therefore independent of the output power. Its present value is two to three times above the theoretical limit

Fig. 11. Last amplifier of ASTERIX III. Aperture 16 cm, active length 8 m. Maximum pulse power 1 TW in 0.3 ns

imposed by diffraction, more than sufficient for the requirements of all laser fusion experiments. The main drawback of the iodine compared with the CO_2 laser is its present efficiency of 0.1%. Although this value is much higher than that of Nd lasers of the same pulse duration and although investigations have shown that an increase of up to 0.5% can be expected in the future, it is not high enough to make iodine a candidate for a power reactor.

It has become clear from the preceding pages that none of the three lasers, neodymium, CO_2, and iodine, comprises all the properties which a reactor laser must have. However, these lasers provide the scientists and engineers with a powerful tool to study the physics and, more and more, also technological aspects of laser fusion. Laser installations in operation today can produce power densities that cover the whole range of interest for a reactor. These installations have opened up the possibility of verifying the results of numerous computer studies which dominated research in the early years of laser fusion. One of the basic questions which has to be answered experimentally before one can predict the feasibility of a reactor is:

6 How Much of the Laser Light is Absorbed in the Target?

It is well known that electromagnetic waves can only propagate in plasmas with electron densities below a critical value where the electron plasma frequency ω_p equals the frequency of the electromagnetic wave ν_L. The plasma frequency ω_p already introduced in the 1920's by Langmuir is a sort of eigenfrequency of the electron gas collectively oscillating around an equilibrium position defined by the much heavier ions:

$$\omega_p = \left(\frac{2\pi e^2 n_e}{m_e}\right)^{\frac{1}{2}}$$

where n_e is the electron density, and e and m_e are the charge and the mass of an electron, respectively.

If an electromagnetic wave is coming from the vacuum into a plasma with increasing density, the wave can only propagate in the region below the critical density. At the critical density the wave is reflected back (Fig.12). In the case of 1 µm light (Nd laser) the critical density is 10^{21} cm^{-3}, in the case of 10 µm light (CO_2 laser) it is 10^{19} cm^{-3}. Comparing these values with the density of solid hydrogen (5×10^{22} cm^{-3}), one immediately realizes that plasmas of solid-state densities cannot be directly heated by laser radiation. Laser light can only penetrate into the outer "underdense" regions of the plasma.. Absorption mechanisms are therefore only effective at and below the critical density.

The main absorption process well known from high-frequency discharges is called "inverse bremsstrahlung" in our regime: The electrons gain oscillation energy from the electric field of the light wave and transfer it to the ions by collisions. This process does not directly depend on the intensity of the light. However, the number of collisions between electrons and ions decreases with increasing temperature. As the temperature of a plasma grows with higher laser intensities, inverse bremsstrahlung absorption becomes less effective and other mechanisms may become more important at higher intensities.

One of these mechanisms is resonance absorption, a process already well known from the propagation of radio waves in the ionosphere. This process occurs at oblique incidence of a p-polarized[4] light beam into a plasma with a density gradient. In this case the electric field vector of the light excities Langmuir waves in the plasma. These waves are then damped, thus heating the plasma. Up to 50% of the incident light energy can be transferred to the plasma in this way. As resonance absorption does not depend on intensity and temperature, this process becomes dominant in the high-temperature regime where collisional absorption has decreased. Resonance absorption has been identified in many experiments. It is seen most clearly in plasmas

---

4 p-polarization = electric field of e.m. wave in the plane of incidence

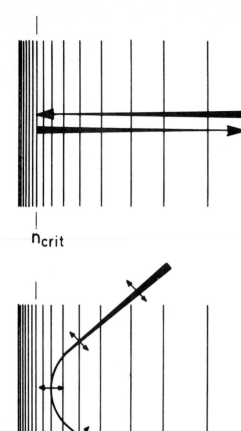

Fig. 12a,b. Light propagation in a plasma with a density gradient. (a) Perpendicular incidence: the light wave only propagates in the lower density region of the plasma and is reflected at the critical density. (b) Oblique incidence and p-polarization: The components of the electric field parallel to the density gradient excite Langmuir waves in the plasma, resulting in additional (non-collisional) absorption

n_{crit}

n_{crit} E

with steep density gradients where the processes in the vicinity of the region of critical density dominate.

However, there are not only favorable processes which help the light to be absorbed. A great number of new phenomena are predicted to occur above certain thresholds of the intensity, some of a good nature, i.e., providing additional absorption, but some of them dangerous because they create additional reflection or scattering losses. As an example of such an "instability", as these phenomena are referred to, we describe here "stimulated Brillouin backscattering" (SBS). This effect can be expected in plasmas with moderate density gradients where the phenomenon in the plasma volume in front of the critical layer becomes more prominent.

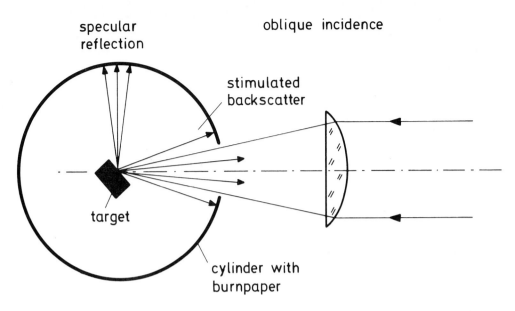

Fig. 13. Experimental arrangement for determination of the angular distribution of the reflected light. A plane target rotatable around an axis is surrounded by a cylinder of photographic paper

The principle mechanism of SBS can be described as follows: Some of the light incident on the plasma is reflected from density structures created by ion sound waves contained in the noise of the plasma and correspondingly Doppler-shifted in frequency. Superposition with the incoming light beams leads to a beat frequency equal to the ion sound frequency and consequently to amplification of the ion sound wave. This, in turn increases the amount of backscattering light with just the right frequency shift for further excitation of the ion sound wave, leading to a further increase of backscattering and so on. If there is no damping of the sound wave then, in the final state, all of the incident light comes back from the plasma without being absorbed, a situation which could be lethal for the realization of a laser fusion reactor. The investigation of SBS in laser plasmas and its contribution to reflection losses therefore touches some of the central questions to be answered.

Here as an example we shall describe experiments performed with the iodine laser at Garching with pulse durations of 0.3 ns and focal diameters on plane targets of up to 400 µm. These are conditions in which volume effects in front of the critical density become important. The experimental arrangement shown in Fig.13 is used to get an overview of the distribution of the reflected and scattered light: The laser energy is concentrated in a small focal spot on a plane target. This target can be turned around an axis perpendicular to the incident beam. It is surrounded by a cylinder made of photographic paper that becomes black by intense radiation. There is one circular hole in the wall of this cylinder giving access to the target for the laser beam. A second hole in the cylinder wall is needed for adjusting the target

HOLE FOR
INC. BEAM

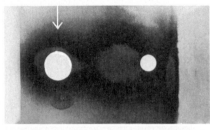

$\varphi_{INC} = 30°$

E$_{INC}$ = 95 JOULE
660 µm, DEF.
NO: 705

Fig. 14. Photographs of
unfolded paper cylinders
for shots with different
angles of incidence in-
dicate backscattering
(into the focusing lens)
and specular reflection

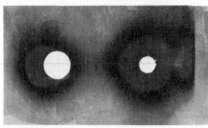

$\varphi_{INC} = 45°$

E$_{INC}$ = 130 JOULE
1060 µm, DEF.
NO: 723

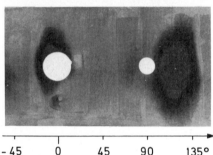

$\varphi_{INC} = 60°$

E$_{INC}$ = 85 JOULE
650 µm, DEF.
NO: 712

-45   0   45   90   135°

relative to the laser beam. The burn pattern produced on the
paper at different angles of incidence of the laser on the target gives information on the distribution of the scattered light.
Examples of such burn patterns at different angles of incidence
are shown in Fig.14. One stain in each picture indicates reflection in the specular direction; the other, surrounding the en-
trance hole, proves that collimated backscattering occurs.

In order to measure quantitatively the amount of light scattered
in the two directions, one photodiode was placed to collect all
the specular light, another to collect the backscattered light.
The results are shown in Fig.15 as a function of the intensity
in the focal spot: The part of the light which is specularly reflected stays essentially constant when the intensity varies
about 4 orders of magnitude. For the backscattered light, however, there is a threshold followed by an exponential increase
and saturation. This behavior and some additional features indicate that Brillouin backscattering is observed. In view of the
prospects of laser fusion as a useful energy source it is very
important that Brillouin backscattering saturates at values of
20% to 30% of the incident energy in contrast to theoretical
estimates for a constant density case which predict backscatter-

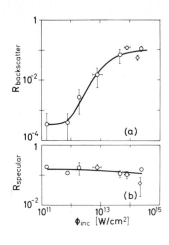

Fig. 15. Ratio of the light energy backscattered (*upper diagram*) and specularly reflected (*lower diagram*) as a function of the laser intensity on target (iodine laser 0.3 ns, spot diameter 400 μm)

ing losses of close to 100%. Although no final theory exists, it is supposed that the fortunately low saturation value is a consequence of the inhomogeneous density profile, in which the matching condition for the excitation of the ion sound waves is only met at very short distances.

In conclusion, these and numerous other experiments have shown that sufficiently high, i.e., between 30% and 50%, absorption can be reached with light of about 1 μm wavelength. Energy absorptions of more than 80% have been observed in the ultraviolet regime. UV lasers or frequency upconverting methods for IR lasers have therefore again attracted interest in the last years. There is another reason for preferring short-wavelength lasers: Resonance absorption is connected with the production of a group of hot electrons with energies well above that of the bulk of the thermal electrons. These hot electrons might preheat the fuel and thus make compression more difficult. It is theoretically predicted and experimentally observed that with shorter wavelength lasers the number of fast electrons is reduced, essentially because the absorption processes occur in a region of higher density where collision processes are of major importance.

Absorption of the laser light is only the first step toward the ignition of the fuel and production of an excess of fusion energy. It must be followed by an efficient hydrodynamic compression process. We shall therefore now discuss the question:

7 Can the High Densities and Temperatures Necessary for Ignition Be Achieved?

At the beginning of laser fusion research no lasers with sufficient energy for compression experiments were available. Only "computer experiments" could therefore be performed. Many codes were developed for this purpose. The most elaborate and sophisticated of these codes is LASNEX, developed and used at the Lawrence Livermore Laboratory. Prediction of these codes indicated that compressions of up to 10^4 times solid-state density could be achieved with laser energies which were thought to be realiz-

Fig. 16. Microphotograph of an "exploding pusher" pellet: A hollow glass sphere of typically 100 μm diameter and 1 μm wall thickness filled with up to 100 atm of DT gas (Courtesy of Lawrence Livermore Laboratory)

able in the future. However, as a great number of assumptions were to be used in the computations, the prediction of the codes involved a wide margin on uncertainty. These uncertainties have been substantially reduced by adjustment to experiments performed in the last few years.

First compression experiments were performed in 1975 using hollow spheres of glass filled with DT at pressures of 50 to 100 atm (Fig.16). Two laser beams from opposite sides were concentrated on such a 100 μm diameter pellet by means of large-aperture optics (mirrors or lenses), resulting in a close to spherically symmetric irradiation. From X-ray pinhole photographs (Fig.17) the compression ratio could be estimated. Maximum values of ten times solid-state density (100 times the original density of the DT gas) were achieved. In such experiments more than 10^{10} neutrons from fusion reactions were observed. A detailed analysis of the results showed that experiments with short pulses (100 ps) and glass spheres with 1 μm wall thickness as used here would not lead to higher compression ratios and finally to ignition. In this mode of operation the thin wall of the pellet is more or less instantaneously heated by the short laser pulse and then explodes radially outward and inward. This leads to a shock wave traveling toward the center, heating and compressing

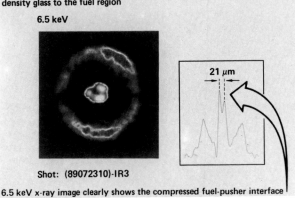

**X-RAY IMAGE DEMONSTRATES FUEL COMPRESSION TO 10×
LIQUID DT DENSITY**

Images at 'high' energy (5 – 10 keV) are required to see through the high
density glass to the fuel region

6.5 keV

21 μm

Shot: (89072310)-IR3

6.5 keV x-ray image clearly shows the compressed fuel-pusher interface

Fig. 17. X-ray image of
an exploding pusher pel-
let (*left*) and densitogram
of the photographic plate
(*right*) (Courtesy of Lawrence Livermore Laboratory)

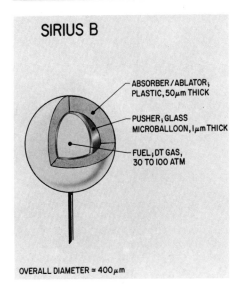

SIRIUS B

ABSORBER/ABLATOR,
PLASTIC, 50μm THICK

PUSHER, GLASS
MICROBALLOON, 1μm THICK

FUEL, DT GAS,
30 TO 100 ATM

OVERALL DIAMETER ≈ 400μm

Fig. 18. Pellet with an additional
outer plastic layer, the ablator, designed to obtain maximum compression
of DT fuel with the HELIOS CO_2 laser
at Los Alamos. In 1979 fuel densities
about 20 times that of solid DT were
reached with this target (Courtesy of
Los Alamos Scientific Laboratory)

the DT gas. However, as this shock wave is far from being isentropic, the density is limited, but relatively high temperatures
are reached.

In order to come closer to the isentropic case and to get higher
densities, one has to change from the "exploding pusher mode"
just described to "ablatively driven" compression. Pellets with
an additional outer layer, the ablator, have to be used (Fig.18).
Programed laser pulses of ns duration evaporate the ablator in
a more controlled manner, resulting in "softer" compression closer
to the isentropic case and leading to higher densities.

Experiments in the ablative mode have been performed with the
SHIVA and HELIOS large laser installations in the United States
since 1978. Final densities of 100 times solid-state density

have been achieved. This is only one order of magnitude below the lower limit of those values necessary for ignition. However, one should mention that the temperatures at maximum compression are lower than in the exploding pusher case. This is indicated by a smaller neutron output observed in these experiments: The energy available in the laser installations is still not yet sufficient to get both high density and high temperature at the same time. This is expected to be possible with the installations SHIVA NOVA and ANTARES now under construction and scheduled for completion in 1984. In the meantime experiments will be continued with the lasers now available. Systematic variations of the various parameters and comparison of the results with computer simulations will lead to a better understanding of the basic phenomena. Especially a detailed comparison of the experiments with CO_2 and with Nd lasers will give information on the influence of the wavelength on the efficiency of light-plasma coupling and compression. In addition, the problems of hydrodynamic stability of the compression process will be experimentally investigated as a complement to theoretical investigations of the growth rates of Rayleigh-Taylor instabilities.

It is expected that "scientific feasibility" of laser fusion will be demonstrated in the middle or the second half of the 1980's. In this context "scientific feasibility" means that the fusion energy produced in an experiment equals the absorbed laser energy needed for ignition. However, this milestone does not yet demonstrate that a laser fusion reactor is technologically and economically feasible. A number of major steps have to follow before this goal is reached.

8 What Will a Laser-fusion Reactor Look Like?

The development of an efficient, high repetition rate, reliable driver is one of the most important reactor problems that has to be tackled. From the technological point of view, of all present lasers only CO_2 could be scaled up to reactor size. However, its long wavelength seems to make compression physics unfavorable. The search for suitable short wavelength lasers is therefore still under way. Out of a number of potential candidates the KrF laser at 250 nm wavelength is favorable as the most promising system with an expected overall efficiency of 5%. But energetic particle beams are also considered as drivers for the reactor. A highly developed accelerator technology is available and large-scale facilities operate reliable with high repetition rates. Although none of these facilities meets all the requirements necessary for the reactor driver, the specialists do not see any fundamental obstacles to the development of an accelerator with the specifications for a net energy producing power plant.

Besides the driver, the other major component of an inertial fusion plant is the reactor chamber. Although there are still a great number of uncertainties in the kind of the driver as well in the physics of the pellet, one has to work on conceptual designs in order to identify the technological problems which have to be solved and devise solutions for them.

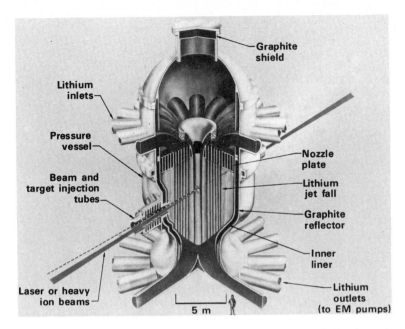

Fig. 19. Reactor chamber of the HYLIFE conceptual design. An array of liquid lithium jets absorbs the energy produced in each ignition and shields the structural wall (Courtesy Lawrence Livermore Laboratory)

One of the most serious technical problems in any type of fusion reactor is to protect the first structural wall of the reaction chamber against damage caused by radiation, energetic debris and neutrons from the thermonuclear reactions. An interesting approach, typical of an inertial confinement system is the HYLIFE fluid wall concept, illustrated in Fig.19. The chamber in this concept is basically made of steel. A blanket of liquid lithium shields the steel walls from the X-ray and neutron energy. The lithium energy-conversion blanket consists of a dense array of jets that are injected into the chamber. The array provides an effective blanket thickness of about 1 m, which reduces the neutron flux to the steel wall by 2 orders of magnitude. In addition, this effectively absorbs the X-rays, the energetic debris and the shocks created by the microexplosion. After each shot the jet array is reestablished in a short time (<1 s) and is ready for the next cycle. The studies have shown that the lifetime of the steel wall protected in this way would be sufficiently long for a practical reactor (30 years, assuming a cylindrical vessel 10 m in diameter and a thermal power of approximately 3 GW). In this design two opposing clusters of laser beams ignite the pellet, which is injected horizontally into the center of the chamber. Instead of lasers energetic ion beams could also be used for pellet ignition.

Although the HYLIFE design is already the second iteration of the fluid wall concept, it is still an early step which will have to be followed by many others as more and more experimental knowledge becomes available in due course.

9 Where Is Research on Laser Fusion Performed?

The largest and most complete program on inertial confinement
fusion is being conducted in the United States. A total of 195
million U.S. dollars is spent by DOE in the fiscal year 1980,
this being approximately 40% of the budget for magnetic confinement fusion (MCF). Most of the money is used for laser fusion in
the National Laboratories of Livermore and Los Alamos. About 5%
is spent for investigations of particle beam fusion.

The effort in the U.S.S.R. on ICF is of the same order of magnitude, being concentrated in the Lebedev Institute and the Kurchatov Institute of Atomic Energy in Moscow. Somewhat smaller but
still substantial laser fusion programs are being conducted in
France and Japan.

All fusion research in the Federal Republic of Germany is part
of a common European program which has traditionally concentrated
on magnetic confinement. Laser fusion is only being pursued in a
relatively small program (about 5% of the fusion budget) by Max-
Planck-Institut für Quantenoptik (MPQ) at Garching. The research
at MPQ is concentrated on a few key problems of laser fusion,
including light-plasma interaction and energy transport at high
intensities and the development of high power lasers for these
investigations.

Since 1978 a small program for exploring the potential of heavy
ion fusion is being conducted by Gesellschaft für Schwerionen-
forschung at Darmstadt (GSI) together with a number of groups
at research institutes and universities. Some exploratory work
was recently started in the field of light ion beam fusion at
Kernforschungszentrum Karlsruhe.

Compared with the large United States and U.S.S.R. programs,
the ICF activities in the FRG are small. However, they might
serve as a basis for a full-scale program in the future when
the progress achieved calls for a stronger effort in this field.

10 When Will Laser Fusion Significantly Contribute to the World's Energy Needs?

This is a very difficult question to answer today since the
feasibility of any fusion reactor has not yet been demonstrated.
Nevertheless, one can try some forecasts· on the basis of present
knowledge. This is done in, for example, the United States fusion
program which is the most comprehensive in the world. In this
program the decision on an engineering test facility (ETF) is
envisaged in the second half of the 80s. This ETF should have
either a laser or a particle beam as a driver. All subsystems
necessary for reactor are integrated in ETF and it should serve
as a test facility for the components of the next step, the
engineering prototype reactor (ERP), which already feeds its
energy output into the public network without necessarily being
an economic proposition. This latter requirement should be met
in the last step, the construction of a full-scale (>100 MW)

The water vapor is precipitated back onto the earth's surface in the form of snow and rain.

Of importance now is that a proportion of this precipitation falls on the earth's land surfaces and thus feeds streams and rivers. These running waters flow downward under the influence of gravity and, in fact, almost always toward the oceans. It is ultimately the gravitational energy of these flowing waters which is harnessed or rather converted to electricity through the exploitation of hydropower.

In order to quantify this gravitational energy, a river reach, with neither inflowing or outflowing tributaries, is considered. The gravitational energy of the mass of water inflowing into this river reach, for example per second, is greater in comparison to that of the equivalent outflowing water mass by the following factor: flow per second times the height of fall (head) times the water density times the earth's gravitational acceleration. The head is practically the difference between the water surface levels at the upper and lower ends of the river reach.

The theoretically recuperable gravitational energy of the considered reach is thus for practical purposes a function of the flow (or discharge) and the head. The water density and the gravitational acceleration can be considered as constants.

What happens to this available gravitational energy if the river reach is not exploited by a hydroelectric power station? It is wasted, i.e., it is "consumed" or rather converted into heat along the reach due to the hydraulic friction (Fig.1, left). This warms the water, the riverbed, and the surrounding air. Assuming, that the energy goes completely into warming the water, this would mean a temperature rise of about 0.24°C per 100 m of height of fall.

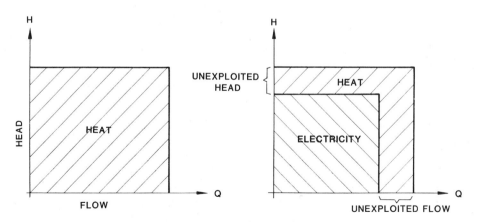

Fig. 1. The hydraulic energy of a river is principally a product of the flow (discharge) Q and the head H. In an unexploited river the total energy is converted into heat (*left*); in exploited rivers only a percentage (*right*), the remainder being converted into electrical current

What happens, however, in the case of hydroelectrical exploitation? The gravitational energy is partially converted into
electrical energy and the remaining part likewise into heat viz.
the heat losses of turbines, generators, and transformers, as
well as into hydraulic frictional heat in the river water. This
frictional heat is then less than that occurring in the absence
of hydroelectrical exploitation (Fig.1, right). Depending on the
overall efficiency of the power station, the river water can
only be heated up a small proportion of the $0.24^\circ C$ per 100 m
height of fall, the minimum being of the order of $0.04^\circ C$.

From this comparison, it is clear how the gravitational energy
can be exploited. One must attempt to ensure an as complete as
possible conversion into electrical energy. That means, one must
construct a power station which as far as possible:

    utilizes the total flow
    exploits the total height of fall
    excludes the heat losses.

These three demands, of course, can only be partially fulfilled;
there are always losses!

1.3   The Losses

The losses which occur in connection with the exploitation of
a river reach are physically, economically, and ecologically
based, i.e., they occur because:

    they are physically unavoidable

    it is economically not worthwhile to fully exclude them

    ecological demands prevent complete exploitation.

Thus, in particular, the complete discharge of a river reach
cannot be utilized. The corresponding losses arise mainly
through:

    Excess water, which for economical reasons is not turbined.

    Residual water, which for ecological reasons must remain
    in the river reach being exploited.

    Flushing water, which is used for flushing out sediment
    deposits, i.e., bed load, suspended load and floating ma-
    terial.

The excess water arises because it is not worthwhile to construct power station features, such as intakes, water conveyance systems and turbines, for all possible discharge volumes.
In other words, the capacity of the power station is usually
limited. Small and medium discharges can be fully exploited;
greater discharges, especially floods, can be only partially
utilized. The discharge volume over and above the utilizable
discharge, i.e., equivalent to the plant capacity, is thus excess. The determining of the capacity, which is also defined
as the design flow (or discharge), results from an economical
analysis.

The residual water is a loss which is peculiar to the so-called river diversion power stations. These power stations divert the water into a side channel away from the river. Thus, this stretch of the river is prone to drying up in periods when no excess water is available. This can be disadvantageous:

Other water-utilizing installations will be prejudiced.

The transport of solid materials in the river will be reduced.

Inflowing waste water will be inadequately diluted.

Navigation and fishing will experience restrictions.

The existence of river flora and fauna will be endangered.

Thus, river diversion power stations normally never utilize the total flow. This means that, even in periods when no excess water is available, some water remains non-utilized in the river. This non-utilized flow is defined as residual or compensation water.

Similarly, the total head of a river reach cannot be fully utilized. The losses occur primarily as a result of hydraulic friction. The corresponding frictional heat along the reach, as described in Sect.1.2, is reduced where hydroelectrical exploitation exists, however it is not eliminated. As this results in a conversion of gravitational energy, this loss manifests itself for a given discharge as a head loss. It is usual to define as head loss only the height-of-fall loss which occurs in the river reach or in the structural features of the power station complex itself. The frictional or height-of-fall losses which occur in the turbines are considered in the turbine efficiency.

In addition to the losses in flow and head, there are also the losses in the electrical-mechanical equipment of the power station. These are:

the energy conversion losses in the equipment

the electrical energy required for the operartion of the power station itself.

This latter energy requirement generally is of little or no significance. The conversion losses, however, occur in the machine units, which comprise turbines and generators, and also in the transformers. The turbines convert the energy conveyed to them into mechanical energy. During this process, hydraulic friction and also bearing losses occur, the latter manifesting itself as turbine heat losses. The generators, which are usually directly connected to the turbines, in their turn, convert the mechanical energy into electrical energy during which generator heat losses occur.

Finally, the transformers transform the tension (voltage) of the generated current into a tension suitable for transportation. Further heat losses are again inherent.

All these losses occurring during the production of electrical current are considered in the efficiency of the powerhouse. This, for modern machine units, attains values between 85% to 90%. In other words, in the powerhouse of a hydropower station,

85% to 90% of the available energy is converted into electrical energy, or, on the other hand, 10% to 15% is lost as heat.

2 Types of Hydropower Stations

2.1 Diversion and Headwork Power Stations

In Sect.1.3 the relationship between head and hydraulic friction losses was described. These frictional losses in any channel are essentially:

   proportional to the square of the flow velocity

   proportional to the channel length

   proportional to the channel roughness (expressed by a roughness coefficient)

   inversely proportional to the channel dimensions (for example the water depth or channel diameter).

Should these losses, in the context of hydropower exploitation, be minimized, then one must obviously design the power station such that the discharges being utilized:

   flow slowly

   flow over a short distance

   flow through large cross sections.

In order to achieve these conditions, two types of power station have been developed. The diversion power station and the headwork power station (Fig.2). The diversion power station, as its name implies, diverts the flow providing power generation from the river via a headwater conveyance system to the powerhouse. Following utilization, the discharge is led via a tailwater conveyance system back into the river. The efficacy of this type of power station lies mainly in that its water conveyance systems exhibit significantly lower roughnesses than the river itself. Through the short-cutting of long loops in the river, the flow distances are often reduced.

Headwork power stations are usually located in the river and are integrated in the headwork impounding the river. Thus both upstream and downstream flows remain in the river. Accordingly, the efficiency of this type of power station relies on water flowing slowly through large cross sections. The downstream flows are possibly slowed down by deepening of the river.

2.2 High-, Medium- and Low-Head Power Stations

Hydropower stations can be differentiated on the basis of their general layout and also on the utilizable head. In practice, there are three types:

   High-head schemes for heads greater than 100 m

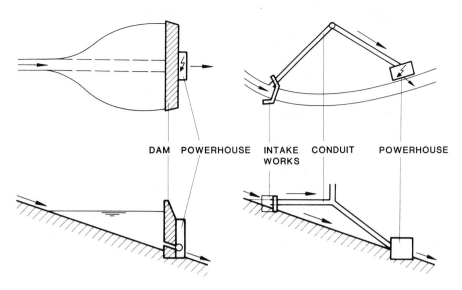

Fig. 2. Types of hydropower station: plan (*above*) and longitudinal section (*below*). Headwork (*left*) and diversion works (*right*)

Medium-head schemes for heads between 30 and 100 m

Low-head schemes for heads less than 30 m.

As the high-head schemes generally occur in the upper reaches of rivers, the exploitable flows are usually small. Conversely, low-head schemes generally are situated in the lower reaches and therefore the utilizable flows are normally large. This means, with regard to their design, that, as a rule, high-head schemes must provide for small discharges at high pressures and low-head schemes for large discharges at low pressures.

2.3 Run-of-River and Storage Schemes

A further possible way of distinguishing hydropower stations is by primarily considering their storage capacities as follows:

run-of-river schemes without storage capacity

storage schemes with storage capacity.

Thus, run-of-river schemes cannot store any useful water; they have to utilize the water available at any given time. There-fore, their energy production varies according to the river flow fluctuations.

Conversely, the storage schemes include storage capacity which is mainly utilized on a seasonal basis. This means that the reservoir will be filled during the "water-rich" season and will be emptied during the dryer season. This results in conservation of excess water until periods of inadequate flows.

The seasonal storage is achieved by impounding valley basins. The impounding works are provided by dams (Sect.4.5).

3 Turbines

3.1 Common Characteristics

As previously mentioned in Sect.1.3, turbines convert the energy of the exploitable water discharge into mechanical energy.
The actual conversion of energy is effected by a "wheel" fitted
with blades or buckets, the so-called runner or rotor. The flow
passes tangentially, radially, or axially to the rotor and thereby causes it to rotate. The turbines thus, from a mechanical
engineering point of view, belong to the turbo machine category.

The turbines are basically divided into two main classes, namely
impulse turbines and reaction turbines. Of these, practically
only three types are now being utilized, i.e., the Pelton,
Francis, and Kaplan turbines. The Pelton turbine or Pelton Wheel,
the runner of which is driven by a free jet of water and which
rotates in the air, belongs to the impulse turbine class. The
Francis and Kaplan turbines, whose runners are subject to the
complete flow of water flowing through them, belong to the reaction turbine class. Turbines, to a certain extent, can be considered as reversed pumps; whereas pumps transfer their mechanical energy into the water flowing through them, turbines do the
opposite, i.e., they extract mechanical energy from the flowing
water. Thus, the Francis turbine is practically a reversed radial pump, the Kaplan turbine a reversed axial pump. This connection is of significance for the so-called pump turbines of
pumped storage schemes.

3.2 The Pelton Turbine (Wheel)

The runner of the Pelton Wheel, as previously mentioned, is
driven by a free jet of water. This flows as indicated in Fig.3
out of a nozzle and jets tangentially onto the runner. There
it impinges on the buckets fixed to the runner, which are shaped
in such a form that they almost completely reflect the jet back
in the opposite direction. In this way, the jet is largely converted into a circumferential force which acts on the runner.
The force transfer is optimal when the circumferential velocity
of the runner corresponds to half the jet velocity.

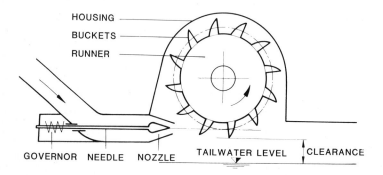

Fig. 3. Longitudi-
nal section through
a Pelton turbine

The flow through the nozzles is regulated by means of a nozzle needle. By the utilization of heads of 400 to 1800 m, the range in which the Pelton turbine is employed today, jet velocities lie between 90 to 190 m/s. The diameter of the jet is limited to about 0.3 m. Large capacity Pelton turbines thus exhibit several jets; there exist versions with up to six jets.

Typical of the Pelton turbine is its suspension clear of the tailwater. The runner must, under all circumstances, be kept clear of the tailwater and is therefore installed at least 1 m above the highest tailwater level. Thus, the corresponding amount of the utilizable head is lost.

The largest Pelton turbine, fabricated to date, is 5 m in diameter and the largest capacity is 300 MW. The best efficiencies are of the order of 91%.

3.3   The Francis and Kaplan Turbines

Whereas the runners of Pelton turbines are only acted upon by individual free jets, the entire flow passes through and completely envelops the runners of the Francis and Kaplan turbines. The forces acting on the blades correspond to the impulse changes of the impelling flows. The Francis and Kaplan turbines are designed, with regard to the flow toward, through, and away from the runner, such that the impulse change, and thus the force transferred, is as large as possible.

In the case of the Francis turbine (Fig.4), the flow into the runner is from tangential to radial and the flow out of the runner axial. The approaching flow passes through a spiral casing which encircles the runner. In this way, the flow is equally distributed over the whole of the runner. The flow exits toward the tailwater via a diffuser which is called the draft tube. The spiral casing and draft tube are essential features of Francis turbines.

It is important to pay attention to the fact that cavitation can occur as the flow exits from the runner. This is the development of vapor bubbles which then collapse. The cause of this phenomenon lies in the redcution of the water pressure below the vapor pressure of water. This cavitation is dangerous as the directly adjacent machinery or structural elements are attacked and pitting is caused which, with time, causes the material surfaces to have the appearance of a sponge. In particular, the runner is endangered by this occurrence. Thus the low pressures arising in its vicinity must be limited such that no, or only insignificant, cavitation takes place.

Because of this reason, the runner cannot be situated at any height above the tailwater level. Placement depends on the type of construction: The runner may be for example only a few meters above or several meters below the lower tailwater level.

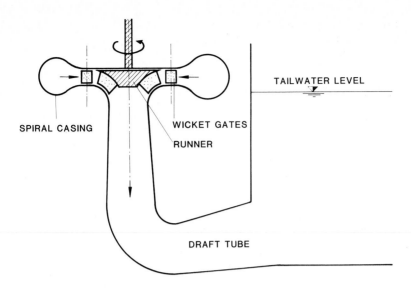

Fig. 4. Longitudinal section through a Francis turbine

Flow through the turbine is regulated by means of a regulating system. This comprises a circular series of adjustable guide vanes called the wicket gates. Francis turbines are used today for the exploitation of heads between 30 to 700 m. Thus, in the 400 to 700 m range, the Francis turbine presents an alternative to the Pelton turbine. The largest fabricated Francis turbines have runner diameters of 9 m (external diameter Gran Coulee turbines 9.9 m). The largest capacities are 750 MW and the maximum efficiencies are of the order of 95%.

In the case of Kaplan turbines, the flow enters the runner axially and also exits axially. As indicated in Fig.5, the runner is rather similar to a propeller and also functions in a similar fashion.

However, a feature of the Kaplan turbine is that the runner is an adjustable propeller. This means that the inclination of the runner blades can be varied to provide optimum conditions for different flows. Therefore, in contrast to the Francis turbine, the runner blades are adjustable propeller blades.

Otherwise, when installed in a vertical axis position, the Kaplan turbine is similar to the Francis turbine. The approaching flow passes likewise through a spiral casing and then via the circle of wicket gates, thus ensuring uniform distribution of flow into the runner. The flow exits via the draft tube, the flow through the turbine being "managed" by the wicket gates. The cavitation problems are similar and thus compel a particular location of the runner with respect to the lowest tailwater levels.

Vertical-axis Kaplan turbines are used today for exploitable heads of 10 to 60 m. The largest versions have diameters of 10 m and the most powerful capacities are 200 MW.

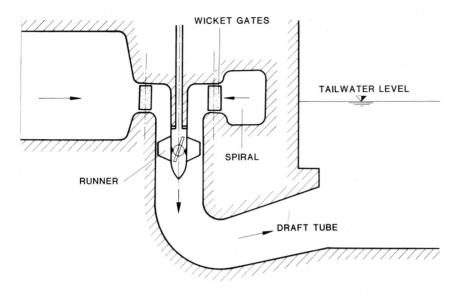

Fig. 5. Longitudinal section through a vertical-axis Kaplan turbine

For low heads, horizontal or inclined axis Kaplan turbines are
suitable. With these turbines, the spiral casing is omitted so
that flows approach the runner more directly and with correspondingly reduced losses. Figure 10 indicates a fairly common
configuration of these so-called tubular turbines.

3.4   The Governor

As already mentioned, the turbines are coupled to the corresponding generators. The coupling is almost always direct, i.e.,
the turbines and generators are located on a common axis. In
modern installations only a few smaller machines (for example,
small tubular turbines) provide an exception. In these cases, a
gearing feature is provided between the turbines and associated
generators.

The turbines and corresponding generators are considered as
machine units. Usually, a single turbine, or more seldom two
turbines, drive a single generator. In either case, the generator produces alternating current of an exactly determined frequency. This means that the speed of rotation of the generator
and hence also that of the corresponding turbine must be kept
constant in accordance with this frequency. This condition is
fulfilled by the action of the governor which reacts to the
slightest variation in the speed of rotation by activating the
necessary regulating features. It thus influences the flow
through the turbine by regulating the nozzle needles in the
case of Pelton turbines and the wicket gates in the case of
Francis and Kaplan turbines.

4   The Main Structural Features of Hydropower Stations

4.1   The Multiplicity

A diversion power station consists basically of at least an in-
take, an upstream water conveyance system, a powerhouse, a downstream water conveyance system, and an outlet. A headwork power
station includes at least a dam or weir and a powerhouse. In
storage power station schemes, a storage reservoir is additional.
These individual features for high, medium, and low head stations
have very different appearances and can be realized in many different forms. The combinations of possibilities of proved designs
are therefore considerable. Thus, it is impossible to describe
here all structural forms of hydropower station schemes. Three
important types have been selected as follows:

   The high-head type (with storage)

   The medium-head type (with storage)

   The low-head run-of-river type.

It is also necessary to emphasize that the structural forms of
a hydropower scheme depend very significantly on their aims and
on the local conditions or characteristics, in particular on:

   the role of the scheme in the supply network (grid) to which
   it is connected

   the hydrological conditions (water availability)

   the topographical characteristics of the location (height
   of fall, length of water conveyance system)

   the geological conditions (quality of site with regard to
   supporting of structures and their construction, etc.).

Hydropower stations are therefore practically always unique and
tailor-made. Whereas it is possible, for example, to construct
the same thermal power station in different locations, this is
practically impossible for a hydropower station.

4.2   High-head Power Stations (with Storage)

This type, as a rule, is realized as a combination of headwork
with diversion works. A storage basin is obtained by the construction of a dam; this being filled with water from the im-
pounded river and perhaps neighboring rivers also (Sect.4.5).
Water from this reservoir is led via an upstream water conveyance system of high pressure conduits to the powerhouse and
from there, via a downstream water conveyance system of low
pressure or free-flow conduit, back to the river system.

It is attempted to keep conduits under high pressure for as
short a way as possible, as these types of conduit are very
expensive. Thus, for high-head storage, two types of construction have come to the forefront. The Alpine system is indicated
in Fig.6 in longitudinal section with exaggerated vertical
scale. Here water is conveyed from the reservoir by a long
pressure tunnel, the power tunnel, to a surge chamber and from

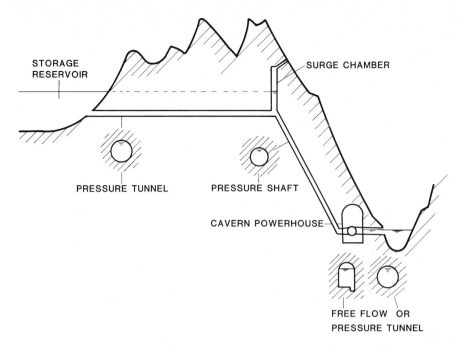

STORAGE RESERVOIR

SURGE CHAMBER

PRESSURE TUNNEL

PRESSURE SHAFT

CAVERN POWERHOUSE

FREE FLOW OR PRESSURE TUNNEL

Fig. 6. Longitudinal section (exaggerated vertical scale) through a high-head power station (with storage) — Alpine system

there by a pressure shaft or penstock, which is kept as short as possible, to a powerhouse either above or below ground. The subsequent tailwater tunnel (tailrace tunnel) or canal is usually short in comparison with the power tunnel.

Under different topographical and geological conditions such as those encountered in Scandinavia, the Scandinavian method is preferred. This is indicated in Fig.7 in a longitudinal section, again with exaggerated scale. The headwater is led by a pressure shaft to an underground power station. Water is returned to the river network by a long tailrace tunnel which is designed either as a pressure or free-flow tunnel. A tailrace surge chamber is usually located at the beginning of a pressure tunnel.

Pressure tunnels are mainly of circular or horseshoe shaped cross section and are lined with concrete if the pressure of the mountain or the internal water pressure warrant it, or if friction losses need to be minimized. Free-flow tunnels have to support only minimal internal water pressures and can thus be more simply executed. Their cross sections are usually portal or horseshoe shaped.

Pressure shafts are either vertical or steeply inclined and are almost always of circular cross section. They are lined with concrete which is often strengthened and sealed with a steel lining. As an alternative to pressure shafts, as in the Alpine type of construction, penstocks can be used. These consist of one or more steel pipelines which are usually laid on the ground surface. The route of the penstocks follows the terrain slopes.

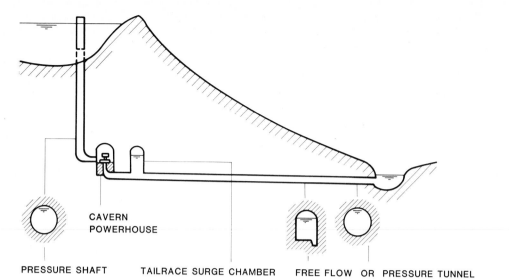

PRESSURE SHAFT   TAILRACE SURGE CHAMBER   FREE FLOW OR PRESSURE TUNNEL

<u>Fig. 7.</u> Longitudinal section (exaggerated vertical scale) through a high-head power station (with storage) — Scandinavian system

Surge chambers mainly serve to reduce the pressure surges occurring in pressure conduits. These surges occur by rapid changes of flow, i.e., by the starting up and shutting down of the turbines and by governor actions. There are several types of surge chamber, the simplest consist, as indicated in Figs.6 and 7, of a vertical shaft. Commonly, in addition to the vertical shaft, an upper and a lower chamber are provided. It is important that the free water surface can rise or sink unhindered. The limiting of the pressure surges causes oscillations of this water surface.

The powerhouses of high-head hydropower stations are equipped with either Pelton or Francis turbines depending on the utilizable head and to some extent on regulating requirements and water quality. Today, in modern installations, Francis turbines are used up to 700 m heads. Above this, Pelton turbines are installed. This dividing line has, with turbine development, been continually pushed upward. The arrangement of the powerhouse depends particularly on the location of the turbine axis and the type of turbine. According to whether the turbine axis (or shaft) is vertically or horizontally disposed, one talks of horizontal or vertical shaft machine units. With the Pelton turbine, as previously mentioned, it is important to ensure that they have sufficient clearance above the tailrance water. This compels an installation of the turbine above the maximum tailwater level, that is to say 1 to 3 m higher. Powerhouses with Pelton turbines are thus situated relatively high with respect to the tailrace levels. This is not the case with Francis turbines which, as a rule, are below the tailwater level and require setting in accordance with the minimum tailwater level. Powerhouses with Francis turbines are located lower, with respect to the tailwater level, than those with Pelton turbines. The size of the powerhouse depends naturally on the size and number of machine units.

Figure 8 indicates an underground power station with horizontal-shaft, single-jet Pelton turbines; Fig.9, a powerhouse equipped with vertical-shaft Francis turbines.

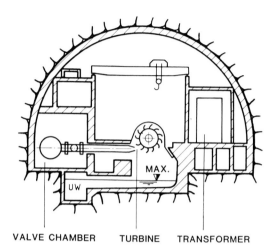

Fig. 8. Cross section through a cavern powerhouse of a high-head scheme with a horizontal-axis Pelton turbine

VALVE CHAMBER   TURBINE   TRANSFORMER

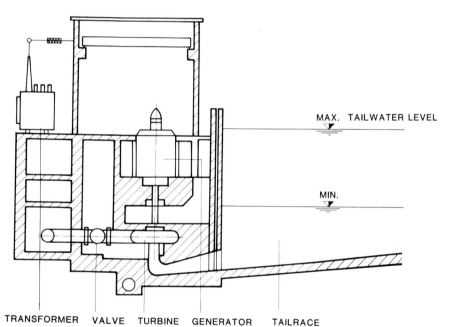

TRANSFORMER   VALVE   TURBINE   GENERATOR   TAILRACE

Fig. 9. Cross section through the powerhouse of a high-head scheme with a vertical-axis Francis turbine

The distribution of the discharge from the pressure shaft or penstock to the individual turbines is effected by means of a manifold. The branches of the manifold are fitted with emergency valves or guard gates. On leaving the turbines, the tailwater is

These ten countries possess with 21,000,000 TJ around 60% of the 35,000,000 TJ estimated total exploitable potential of the earth. In some of these countries up to 1974, and also to some extent to date, only an insignificant percentage has been developed.

Table 4. The ten countries with the largest exploited hydropower potentials in 1000 TJ (according to Cotillon 1978)

		Exploited potential 1974	Degree of development 1974
1	United States	1,100	44%
2	Canada	760	40%
3	U.S.S.R.	480	12%
4	Japan	300	63%
5	Norway	280	64%
6	Brazil	240	13%
7	Sweden	210	57%
8	France	200	88%
9	Italy	140	80%
10	Spain	110	46%
Total approximately		3,800	

These ten countries provided in 1974 with 3,800,000 TJ around 73% of the total exploited potential of the earth at that time of 5,200,000 TJ. Several countries in Europe as well as Japan had already achieved a high degree of development and in the meantime have approached the stage of final development of their hydropower potentials.

It is interesting to note that the contribution of hydropower to overall electricity generation for the earth as a whole has sunk from about 23% in 1974 to a little less than 20% today.

Table 5. The ten countries with the greatest proportion of hydropower in their overall electricity supply networks (according to Cotillon 1978)

Rank	Country	Hydropower proportion 1974
1	Norway	100%
2	Zaire	96%
3	Brazil	95%
4	Zimbabwe	90%
5	New Zealand	84%
6	Switzerland	77%
7	Sweden	76%
8	Canada	75%
9	Columbia	75%
10	Portugal	74%
	Earth	23%

It may be concluded that worldwide efforts are being made to
develop hydropower exploitation. In some lands still great unused possibilities lie awaiting development. The evidence of the
growing scarcity of fossil fuels for thermal power stations is
causing many countries to revalue their hydropower potential
and to take positive steps to ensure their exploitation. Certainly, some hydropower projects are remotely located, far from the
energy consuming population centers, so that important energy
transmission problems are presented.

As an example of a large project, the Itaipu Hydropower Project
is mentioned here. This project which is at an advanced stage
of construction has an installed capacity of 12,600 MW and is
to date the largest hydropower project in the world. It lies on
the border between Brazil and Paraguay on the river Parana, one
of the largest rivers of the earth and has the following main
characteristics:

Reservoir	Volume	29 billion m3
	Surface area	1,350 km2
	Length	151 km
Dam	Length	8 km
	Max. height	190 m
Powerhouse	Design flow	12,600 m3/s
	Max. head	130 m
	Installed Capacity	12,600 MW
	No. of machine units	20
	Average annual energy	18 TJ
Main quantities:	Soft excavation	21 million m3
	Rock excavation	25 million m3
	Concrete	12 million m3

6 Pumped Storage Schemes

6.1 The Aims of Pumped Storage

It was already mentioned in Sect.5.1 that electrical energy as
such cannot be stored. If an excess is available, it must be converted into another energy form and later reconverted back into
electrical energy. This requirement is fulfilled by the use of
pumped storage schemes; they utilize the excess electrical energy to pump water into a higher level reservoir in order that it
may subsequently be exploited. In other words, they convert the
excess electrical energy into gravitational energy in the water
and enable its reconversion in a hydropower station.

Pure pumped storage schemes essentially consist of upper and
lower reservoirs with connecting water conveyance systems and
a power station (Fig.20). As a rule, water is pumped during the
night and weekend from the lower into the upper reservoir so that
on returning by the same route during peak demand periods on
weekdays it may be turbined. The power station is thus equipped

with both pumps and turbines. Naturally, this cyclic operation is subject to energy losses.

The most efficient pumped storage schemes require about 1·3 J of energy for pumping for each Joule of energy produced by the turbines, i.e., the overall efficiency of the plant reaches a maximum of about 75%. A pure pumped storage scheme is thus in fact not really a power station; it does not utilize any natural resource to generate electricity. Hence the particular omission from Sects.1-5 and the inclusion herein as a separate section. It is similar in almost all details to a hydropower station, but does not contribute to the electrical energy supply situation, rather solely to the regulating of the supply to correspond with the demand, i.e., in principle it transfers electrical energy from periods of excess supply to periods of scarce supply.

Where water flows into the reservoir from the immediate or neighboring catchment areas, this is turbined together with the storage water. Such an installation is no longer defined as a pure pumped storage scheme, but rather as a pumped storage scheme with natural inflows or as a hydropower station with pumps. In reality, the only difference here is that the powerhouse is not only equipped with turbines, but also with pumps. In this case, it not only serves as an energy generating facility but also as an energy transfer facility. The multiplicity of existing high-head schemes includes therefore all transitional features between pure pumped storage schemes and high-head schemes.

6.2 Design Details

It was already mentioned in Sect.4.5 that the energy stored in a storage reservoir is basically a product of the useful storage capacity and the head. Therefore, pumped storage schemes are concieved in the main as high-head schemes and extensively include similar features (Sect.4.2). The most obvious differences lie in

the construction of the upper reservoir

the presence of a lower reservoir

the combination of pumps with turbines.

The upper reservoir may be formed by the construction of a dam or a so-called artificial reservoir. These latter reservoirs in some pure pumped storage schemes are situated on the tops of mountains. An embankment dam encircles the reservoir area and the inner surfaces are provided with a sealing coat, usually bitumen based. The lower reservoir may also be formed by one of the two above mentioned means but it is also possible to utilize a natural river or lake or even the sea for this purpose.

In flat coastal areas, there exists an interesting proposal in that the sea be used as upper reservoir with the lower reservoir being in the form of a cavern excavated several hundred meters below sea level. Figure 18 indicates a possible arrangement.

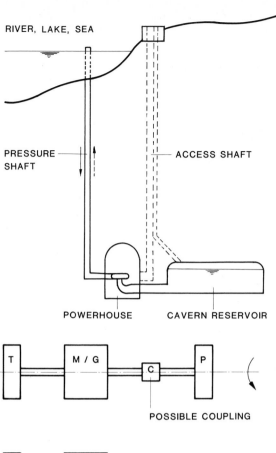

RIVER, LAKE, SEA

PRESSURE SHAFT — ACCESS SHAFT

POWERHOUSE   CAVERN RESERVOIR

Fig. 18. Longitudinal section of a pumped storage scheme. Proposal for the utilization of a river, lake or sea as upper reservoir

POSSIBLE COUPLING

Fig. 19. Machine units of a pumped storage scheme. *Above*: scheme of a three-component maschine with a Pelton turbine. *Below*: scheme of a two-component machine

As previously mentioned, the powerhouse of a pumped storage scheme includes pumps in addition to the turbines. Usually, these are combined. Figure 19 demonstrates two possibilities. *Above*, a so-called three-component machine unit is shown. Here, the turbine and pump are both located on a common shaft together with the generator. During turbining, the generator-motor works as an electrical current generator, and when pumping as an impelling facility. By means of the coupling feature, the necessity of emptying the pump housing of water when switching over to turbine operation can be avoided. *Below*, a so-called two-component machine is indicated in which the generator-motor is combined with a pump-turbine. The pump-turbine is essentially a radial pump which, by reversing the sense of rotation, acts as a Francis turbine. The runner serves as a runner for both pump and turbine. This machine is also known a reversible machine.

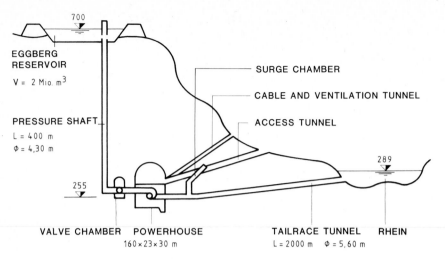

Fig. 20. Longitudinal section through the Hotzenwald pumped storage scheme near Säckingen, West Germany

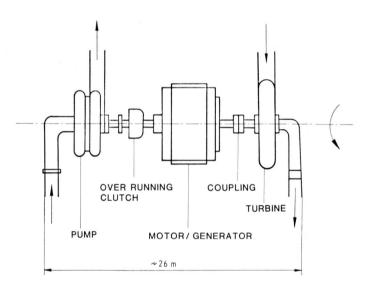

Fig. 21. Three-component machine of the Hotzenwald pumped storage scheme (ref. Fig.20)

The selection of machine type depends on the head, discharge volumes, and the structural and operational conditions. It is clear that a two-component machine is less expensive than a three-component one and in particular requires less space. However, the change over from pump to turbine operation or the reverse is slower because the sense of rotation has to be reversed, as with reversible pump-turbines. Appropriately equipped three-component machines effect this changeover within less than a minute, reversible machines in 5 to 10 min or with more expensively designed versions between 1 1/2 to 3 min.

Figures 20 and 21 show, as an example, the Hotzenwald Pumped
Storage Scheme near Säckingen in West Germany. This scheme in-
corporates between the upper and lower reservoirs a head of
410 m and has the following capacities:

pumping: discharge 64 m3/s, capacity 270 MW

turbining: discharge 96 m3/s, capacity 330 MW

upper reservoir volume: 2 million m3.

Acknowledgements. I would like to thank Motor-Columbus Consulting Engineers
Inc, Baden, Switzerland, for their friendly stimulation and suggestion and
in particular for the translation into English.

Solar Power Plants[1]

H. Treiber[2]

1 Introduction

Fossil fuels are not inexhaustible. When this, due to the so-
called oil-crisis, was realized by large sections of the public,
the search for alternative sources of energy began; alternative
also to nuclear energy, the qualities of which were not incon-
testable. The most promising candidate was solar energy, which
has already long been used in its subordinate forms: timber,
wind, and hydropower. The direct exploitation of commercial
energy from solar radiation through power plants is, however,
more problematic than is generally assumed (Fig.1).

The Earth's surface is exposed to solar radiation with a rather
low energy density of not more than a maximum of 1000 W/m^2, a
rate very seldom really reached. In our hemisphere the medium
rate lies at roughly 500 - 600 W/m^2. To be thermally applicable
this energy first has to be "densified" before it can be transformed.

The energy reaches us in form of light radiation, that is partly
dispersed when passing through the atmosphere so that, depending upon the weather conditions and the climatic zone, a mixture
of diffuse and direct radiation is available. Diffuse light
cannot be concentrated optically, i.e., through converging lenses
or burning reflectors. This means that, depending upon the location, certain types of power stations cannot be taken into
consideration.

Since the sun does not shine 24 h, a solar plant can, roughly
spoken, operate only half a day. The rest of the time it stands
unused compared to a conventional power plant that is not subjected to such limitations. The angle of incidence of direct
solar radiation depends upon the movement of the sun and upon
the geographic position of the site. The angle of radiation becomes more obtuse the further north the site is located, meaning
that higher losses must be put up with when collecting the light.
By applying the appropriate reflector systems these losses can
be kept within limits, since, e.g., a concave reflector due to

1 This paper was translated by Mrs. S. Messele-Wieser

2 Messerschmidt-Bolkow-Blohm, MBB Raumfahrt, Postfach 801169,
  8000 München 80, FRG

253

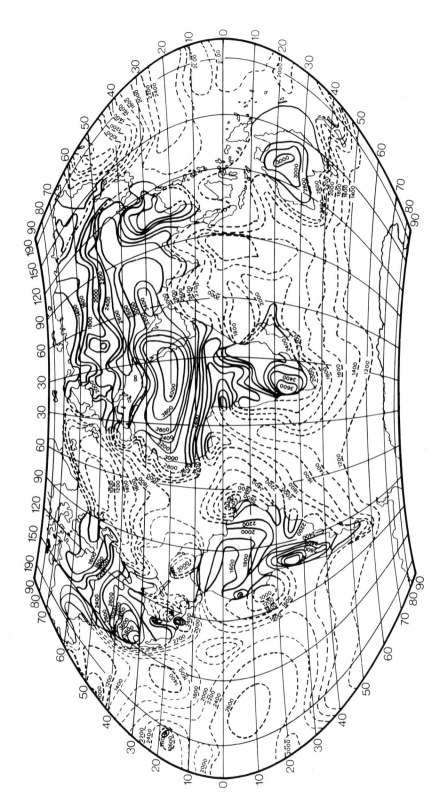

Fig. 1. Average annual insolation in hours

Fig. 6. Kuwait project (Industrial photo MBB)

considering the technical complexity of such an integrated system, there is still a long way to go. These required standards
should not unnecessarily raise the cost of the plant, since
this would impede economic efficiency.

6   Stirling Solar Power Plants

The problems as mentioned above with operating media, heat transfer media, operating equipment, lengths of tubes, losses during
circulation, etc., are at once eliminated if the heated side of
a stirling cycle engine is set in the focus of a concentrator.
The quality of this reciprocating motor lies in the fact that
any gas with good thermal conducting properties, like, e.g.,
helium, expands in a piston when heated from any source of heat,
generates energy while cooling off, and can then be reheated
again.

Solar energy is ideal for this, since it can be concentrated to
extremely high temperatures and is completely free of impurities
that could cause problems on the radiation receivers. The trouble
is that a dependable large-scale operational stirling cycle engine is not yet available. This situation may soon change, however, under the pressure of solar energy developments.

7   Tracking of Farming Plants

Concentrating plants operate with directed radiation. Since the
angle of incidence changes with the position of the sun, the reflector systems have to track the sun. Since the individual receivers are fixed to the reflectors the focus may, in principle,

move the reflectors which are turned in such a way that they are
fully facing the sun, meaning that the focus is on the symmetri-
cal line and this is pointing directly to the sun. If the receivers are to be moved, the cables from and to the receiver
will have to be flexible — this is not really a minor technical
problem when dealing with a temperature of 300°C and even higher
pressures.

These difficulties can be eliminated by constructing the reflector around the fixed receivers in such a way that the focus remains stationary. In this way only the direction of incidence
changes, i.e., the radiation receiver must be able to convert
the radiation coming from different directions within a certain
angle. The parabolic reflector in Fig.5 is constructed on this
principle. Tracking is done either with the help of electromo-
tors (strongly reduced in size) on one axis, the other being inclined like an astronomic telescope, corresponding to the geographic latitude and the time of the year. It is only adjusted
by hand once a week. Because of low requirements as to exactness,
motors, gearing and regulating electronics can be relatively
simple and therefore cheap. This is also the case when both axes
are moved automatically.

8 Tower Power Plants

It has already been mentioned that farms need collecting pipes
to collect the heat transfer medium from the individual receivers
distributed over a large area. Among others, this factor determines the field size for maximum efficiency. When changing over
to a higher output solar energy is not therefore first transformed
into thermal energy and then collected, but instead the collecting takes place at the stage of light radiation, so that the
pipe system and the additional costs and losses resulting from
it are eliminated.

One scheme is shown in Fig.7. Heliostats are distributed over a
field. In the south of the field there is a tower with a receiver on top. There is only one receiver for the whole field,
meaning one common focus for all the heliostats. Radiation energy is transformed into heat within the radiation receiver and
then conveyed further with the help of a heat transfer medium.
The pipe system consists of the distance between the tip of the
tower to the evaporator and the engine (independent of the field
size). If both are set within the tower, this distance can be
kept very short. (Fig.7 see next page)

9 Heliostat Field

A heliostat (Fig.8) consists of a supporting structure, the individual reflectors attached to it, the gearing, the motors
and an electronic unit. Heliostats track the sun in two axes.
Considering the sometimes enormous distance from the heliostat
(if it is situated at the edge of a field) to the tower (300 m)
and in view of the high focal temperatures on the receiver, it

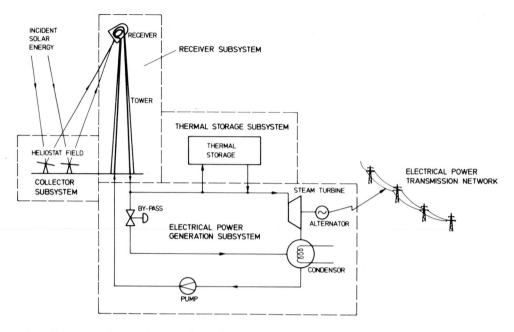

Fig. 7. Central receiver solar thermal power system

Fig. 8. Heliostat (Industrial photo MBB)

is understandable that, if only for security reasons, the focus
must be very precisely positioned. This is not only necessary
in order to be able to emit as much energy as possible to the
receiver, but also to ensure that the receiver does not start
burning at the fringes. Here a tracking exactness of less than
10 s of arc is necessary. This means that the whole structure
must be very stable in the direction of radiation and vertically
to the reflectors, so that the focus does not start moving at
normal wind force. For the same reason, the gearings are also
very stiff, not only do they have no backlash, but the pinions
and cogs must also be very stiff. Heliostat tracking first com-
menced in a way that the reflected ray of a small reference reflector was "seen" by a sensor standing between the heliostat
and the tower. If the point of light moved away from the center
of the sensor because, e.g., the sun changed position, the sensor transmitted the corresponding signals to the control electronics, which then set the adjusting motor into motion. Tracking therefore proceeded in a closed loop. The system seems to
be simple, suitable microprocessors or even calculators were not
known then. Yet the system does have grave disadvantages. At
sunrise already, when the plant was started up, the heliostat
had no information where the sun was to be found. If the sun was
obscured by a cloud, the heliostat "lost sight" of the sun.
Finally, this system tended toward dynamic instability, since
within the closed loop the structural stiffness has a certain
influence that may lead to a constant foreward and backward movement of the motors, which could be checked but only at the cost
of tracking accuracy.

Developments in processor techniques have, by now, made a completely different tracking mechanism possible that basically
functions as follows: A central calculator regularly feeds the
coordinates of the sun's position according to time and season,
into a distribution network. The heliostats are connected to
this by their own microprocessors. There the data about the sun
and the data of the heliostats' position in the field in relation
to the tower receiver, meaning the focus, are processed into
angle information for the two axes of the reflector and then
fed into the drive mechanism as nominal value. The adjusting
mechanism then turns the heliostats until the received nominal
angle is reached so that the sun is projected in the opening of
the receiver. This procedure is repeated several times each
minute.

The tracking principle described above is called an open loop.
Compared to the closed loop it has considerable advantages. In
a closed loop each heliostat sets the adjusting mechanism in
motion when, metaphorically speaking, the individual regulators
felt like it, which could lead to the result that almost all adjusting motors of a whole field were working at once and consuming electricity accordingly. In an open loop, however, the individual heliostats are directed group-wise so that the total energy consumption of the whole field remains level without extreme
peaks. The ability of always "knowing" where the sun is standing, even if it is not visible or hidden behind a cloud, or in
the morning when the reflectors are brought out of their neutral
position, is one of the most important advantages of the open

loop. In addition, such a system cannot make such mistakes as has allegedly happened in a closed loop, where a small but very luminous cloud was mistaken for the sun and tracked by the heliostat.

Besides the main task of tracking, the heliostat's control system has also other important functions so as to guarantee a reliable service of the solar power plant. In case of breakdowns in the thermic part of the power plant, the heliostat naturally has to be moved out of focus to keep the receiver from burning out. Or the heliostat has to be put in a neutral position, that is, with the reflector facing downward, be it over night to keep it from becoming moist with dew, or be it during a storm to keep the structure from becoming overloaded. In these cases special orders can be fed over the data distribution network into the central calculator. These orders then have priority over the normal operating orders.

The output of a heliostat, meaning the fraction of solar light collected and reflected in the receiver aperture, is not only dependent upon the stiffness and accuracy of tracking, but also largely dependent upon the optical properties of the mirrors. The mirror itself becomes less stiff, the thinner the glass layer over the actually reflecting surface becomes, so that from a certain thickness of the reflecting glass onward, a supporting structure of expanded plastics, sandwich iron sheets or framework becomes necessary. These additional structural parts naturally add to the total weight, thereby increasing the mechanical problems. A flat mirror would reflect sunlight, but only bring a minor fraction of it into focus, the larger part would pass the receiver by and be lost. For this reason, each mirror must be concave-shaped so that the image of the sun is reflected in the focus (Fig.9) and no energy is lost. The supporting structure of the mirror must meet these requirements. The individual mirrors of a heliostat are then arranged on the heliostat structure in such a way that all sun images of the individual mirrors coincide in the receiver. Adjusting screws are used for adjusting each heliostat so that it is focused. Strictly speaking, this focusing only applies for one certain angle of incidence of the solar radiation. Under other angles the focus is widened according to optical-geometrical principles and the individual images of the sun no longer coincide. This must already be taken into consideration when first laying out the size of the heliostats according to the receiver aperture and the geographic latitude, since even with the focus widened to a maximum it must still be within the receiver aperture.

All these factors that influence the heliostat's efficiency are negatively affected by the principles of stiffness, reflecting properties, and adjustability, since they add to the total weight. An alternative construction would consist in putting the heliostat under a transparent cupola with a protective gas atmosphere. The structure is then protected from wind and can be built very light; there are no more problems caused by weather so that it is possible to utilize metalized plastic sheetings stretched over a frame, as reflectors. In short, the heliostat could be built in an almost ideal way. One disadvantage is the

<u>Fig. 9.</u> Image of sun on target; EURELIOS (Industrial photo MBB)

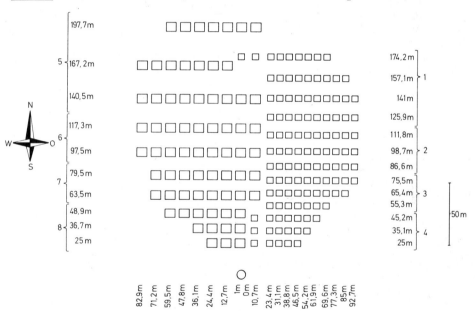

<u>Fig. 10.</u> EEC-1 MW solar power plant. Heliostat field layout. Height of receiver aperture 50 m above ground. Field inclination 4% from south to north

fact that the transparent, pressure-proof cupola which would be needed causes a loss of light and brings additional expenses. Experience will have to show which principle has the best chances in the future.

Hundreds of heliostats are arrayed at the foot of the tower (Fig.10). The closer they are to the receiver, the more heliostats can be set close to the tower. At the same time they are, however, shading and blocking each other so that compromises have to be made concerning the field setup. The degree of blocking and shading varies, depending upon the angle of solar incidence during the day, the time of the year, the geographic latitude, the terrain, and the height of the tower. Finding out the best compromise is so complex that it has to be done by big calculators through costly optimization programs. One usually tries to generate as much energy as possible in one year. It is, however, possible to go by other optimisation criteria such as shorter periods, minimal area needs, or maximum daily output.

For obvious reasons, the field has a symetrical layout from north to south, the tower being located in the south (north-south field, Fig.10). This arrangement has its limits though, since with a growing number of heliostats for larger power plants, the reflectors at the end of the field would be too far off from the tower, meaning additional expenses for increased accuracy demands. A lateral expansion would have the effect that the heliostats on the side of the field would "see" the receiver opening under a far too acute angle, so that on the one hand the reflected light would only graze the receiver, most of it being wasted, on the other hand the thermic load on the receiver would be quite one-sided.

When designing large heliostat fields for an output of more than 10 MWel, it is possible, therefore, to increase the height of the tower, which again has its limits due to increased costs. An effective solution would be placing the tower in the middle of the field (Fig.11). Compared to the north-south layout, the receiver construction is altered so that now two or more apertures are arranged on the tower. The heat exchanger plates inside the receiver are structured accordingly.

Fig. 11. 20 MWe gas-cooled solar tower power plant

10   Thermodynamic Loop

Tower power plants have a high concentration factor, which means
that the area for collecting the incident solar light is very
large in comparison with the receiver area. Flat-plate collectors have a concentration factor of 1 - 5, i.e., there is no or
only very little concentration; parabolic linear reflectors reach
50 - 80; parabolic reflectors reach approximately 200; tower
fields reach 500 - 2000. From a factor of 3 onward no diffuse
light is processed, this influencing the operational capacity
as described above. The operating temperature can be higher; the
higher the concentration factor becomes, thermodynamic effectiv-
ity is higher and the solar power becomes cheaper.

In the heat exchanger of the receiver, incident energy of solar
light is transformed into heat in any case. This heat is then
transferred to a working medium over a heat transfer medium, as
already described for the parabolic reflector farm. Since tower
plants achieve higher processing temperatures (from 450°C to more
than 1200°C), heat transfer and operational media with a basically higher thermal stability must be utilized. The classic
choice is steam. The tower receiver then becomes a radiation
boiler. A superheated steam pipe leads to the turbine and the
solar power plant is nothing more than a solar heated steam
power plant, though with a highly specialized radiation boiler.
More data has to be collected about the construction of this.
In Adrano, Sicily there is such a power plant. The contract for
it was awarded to an international consortium of firms by the
Federal Republic of Germany, France, Italy, and the EEC (Fig.12).

Another version of the solar heat-driven steam power plant em-
ploys metalic sodium, as is being used for the cooling of
valves in engine construction and in nuclear reactor technology,
instead of steam for transporting heat from the receiver to the
vaporizer. The receiver becomes simpler, since no high pressures
develop any longer. A research reactor of this kind is being
built in Almeria, Spain, by the International Energy Agency
(I.E.A.).

When utilizing steam, an operating temperature of more than
550°C leads to high pressures that would increase the expenses
for the whole piping and heat exchanger system, and the turbine,
too, would become unnecessarily expensive because of the extreme
volume differences between entry and exit. Therefore, other
methods must be applied when using processing temperatures above
600°C in order to boost thermodynamic and total efficiency. Gas
turbines operate with higher temperatures. This is taken advantage of in a project that is being examined further in commission of the Federal Republic of Germany. In the receiver of a
tower power plant, gas is superheated up to 800°C and sent through
gas turbines. After emerging from the turbine, the still sufficiently heated gas emits heat to the vaporizer of a simple steam
turbine cycle, before again being conveyed to the receiver and
reheated.

Fig. 12. European solar tower power plant "Eurelios" in Adreano (Sicily)

Besides this closed gas cycle there is another type in which the compressor takes in air, pushes it through the receiver where the air is superheated, and then propels a superheated-air turbine. The turbine, for its part, drives a compressor and the generator. Subsequently the air, either directly or after passing through a heat exchanger, streams outside. This variant is called an open gas cycle. The advantage of simplicity compared to the closed cycle is set off by disadvantages in efficiency under certain operating conditions.

Both gas cycles have a superheated-gas pipeline between receiver exit and turbine entry, that has to be as short as possible because of the extraordinarily high costs. This is why in both cases, and contrary to the types mentioned above, the turbines are not on the ground, but close to the receiver on the tower.

The stirling cycle engine could operate without any outer superheated-gas pipeline, the heated parts can be arranged around the focus of the reflector field. For this reason it is on the tower. Stirling cycle engines are, however, as mentioned still being developed and are therefore not yet available for use in solar power plants.

Because of their higher processing temperatures and lower losses, tower power plants have a high degree of efficiency reaching from 16% to approximately 30%. How much of this can be realized in the first run will be seen in the coming years, when the individual plants now under construction will start experimental operation.

11 Classification of Solar Power Plants

In the pro and contra of the energy dispute, solar power plants as well as coal and nuclear power plants are not always seen objectively since it deals often only with superficial questions connected with the generating of current. Certainly, our energy problems can not be solved by solar energy alone, least of all in our moderate-temperature latitudes. It is likely that solar power plants will have their place in an integrated supply structure of hydraulic power, pumping storage stations, and other future systems depending upon renewable energy sources.

Acknowledgement. The author is indebt-d to the "Fachinformationszentrum Energie-Physik-Mathematik GmbH" in Eggenstein-Leopoldshafen 2 for the immediate and extensive compilation of bibliography references on the subject.

References

Altenpohl D, Mettler DH (1976) Aluminium as structural material for solar radiation collectors. Aluminium, Düsseldorf Mar:184-187

Bammert K, Poesentrup H (1978) Steam and gas turbines for small solar power plants, Atomkernenergie 1978:153-158

Bammert K, Krapp R, Reiter U (1978) Gas turbine cycles for effective use of solar energy, 10th worlds energy conference, 19-24 Sep 1977. Atomkernenergie 1978:1-3

Choinowski H, Oertli E (1979) First solar farm for Europe. Forsch Planen Bauen Apr 1979:53-59

Dornier System GmbH, Friedrichshafen (1979) Small solar power station at Cairo/Egypt, Pamphlet (4 pages)

Electricité de France (EdF), Paris (1980) La centrale solaire expérimentale THEMIS, information Pamphlet (16 pages)

EPRI Journal (June 1979) Turning to the sun for power

Equipment for generating electric energy from solar energy. German (FRG) Patent 2,619,480A/

Equipment for conversion of solar energy into mechanical and electric energy. German (FRG) Patent 2,525,337/A/

Essay on a solar power plant. Frankfurter Allgemeine Zeitung: 1, June 1979

Fanninger G, Gilli PV, Heindl W, Kleinrath H, Korzen W (1979) Solarthermische Kraftwerke, technische und wirtschaftliche Möglichkeiten in Österreich. ASSA (Austrian Society for Solar Energy and Astronautics) May 1979 (detail bibliographical data)

Feustel JE (1978) Development of small solar power systems, presented at the Solar Meeting in Genua, June 15-18,1978
Feustel JE (1979) Performance and cost optimisation of small solar power stations. Beitrag zum ISES-Kongreß vom 28.5.-1.6.1976. Atlanta, USA

Feustel JE, Mayrhofer O, Wiedman U (1979) Solar farm systems, layout application economy efficiency, presented at the ISPRA-Seminar, Solar thermal power generation, September 3-7, 1979

Gerster G (1978) Auf dem Wege zum Sonnen-Großkraftwerk. Neue Züricher Zeitung: 9, December 1978

Glatzel F-J, Stoy B (1978) Zum Einsatz von Sonnenkraftwerken. Elektrizitätswirtschaft, Broschüre 20

Interatom, MAN, MBB, Dornier, KWU, GHH (1980) Gas-cooled central receiver solar power plants, information pamphlet (12 pages)

Hopmann H Development of solar thermal power plants, 1st German solar energy forum, vol II. Proceedings, pp297-313

3rd International Solar Forum (1980) Conference report. Hamburg, FRG, 24, June 1980

Isenberg G (1978) Solar farm power plants of the category 50 to 500 kW, presented at the Internatl Colloq New Sourc Energy, Rabat (Morocco), November 1978

Kleinkauf W (1977) Solar-thermic systems. Mitteilungsbl Dtsch Ges Sonnenenergie Jan

Matthöfer H (ed) (1976) Solar energy, vol I and II. Umschau

Meinhardt H (1977) Can 'solar electricity' be generated economically. Comparison between a fossil-fired and a solar 20 MW steam power plant in a sunny region. Elektr Energ Tech 1977:61- 64

Messerschmitt-Bölkow-Blohm GmbH, Munich (1979) Solar energy technology at MBB, information pamphlet (12 pages)

Messerschmitt-Bölkow-Blohm GmbH, Munich (1979) 100 kW_{el} solar power plant, information pamphlet (20 pages)

Palz W (1978) Solar electricity, an economic approach to solar energy. Butterworth, London Boston, UNESCO

Schwarz K (1978) Considering the operation of solar power plants under the aspect of energy yield. Mitteilungsbl Dtsch Ges Sonnenenergie Jul-Aug

Simon M (1977) Die Zukunft solarthermischer Klein-Kraftwerke, presented at the ASE-Meeting "Sonnenenergie, Anwendung/Systeme/Erfahrungen", Essen

Simon M (1978) Plans for thermal solar power plants. Mitteilungsbl Dtsch Ges Sonnenenergie Jul-Aug

Solar meeting in Bari, Italy, June 1979, report

Solar thermal power plant. German (FRG) Patent 2,553,283/A/

SSPS-Project: Two solar power plants in Spain (1979) Dtsch Forsch Versuchs-anst Luft-Raumfahrt Nachr Nov 1979:15-17

Swiss Solar Energy Society (SSES) (1975) Krafterzeugung mit Sonnenenergie und Langzeitspeicherung. Symposium III, 1, December 1975

Welches Solarkraftwerkskonzept is besser? Frankfurter Allgemeine Zeitung vom 22. April 1980

Electricity from the Sun – Photovoltaics

E.F.Schmidt[1]

1 Introduction

The worldwide activities in research and industry, with considerable state participation aimed at exploiting solar energy to supply electric power, has to be seen within the context of strategies directed to a long-term solution of energy problems. Newly awakened interest in utilizing mankind's oldest source of energy results from its outstanding qualities such as inexhaustibility, positive environmental quality of energy conversion, and decentralized availability. Of the two technical processes as a systematic solution regarding energy conversion and energy storage for economical electricity production, namely

 direct photovoltaic conversion of solar energy into electrical energy in solar cells transferring radiation energy to the charge carrier of a semiconductor,

 indirect conversion of solar energy initially into heat in collectors and then by way of a thermal dynamic process into mechanical and electrical energy,

the present contribution will deal only with the solar cell systems.

At the end of the 1950's solar cell generators proved their worth to an notable degree even under the extreme conditions of outer space and were accepted as a means of providing electrical power in space flight. Thus, also at the early stage investigations were carried out concerning terrestrial power production with these completely static systems. However, it was the major progress in silicon semiconducted techniques which eventually led to worldwide engagement in favor of terrestrial utilizations. This method of energy conversion, by far the most elegant and simple, thus for the first time in the energy technique sector offers prospects of being competitive in all power ranges and, therefore, of being a genuine alternative as a decentralized electricity supply of the future. Beyond the Earth's atmosphere (AMO = Air Mass O) the energy from solar radiation indicates with sufficient accuracy the spectral distribution corresponding to the radiation of a black body with a temperature of 5900°C. The intensity of solar radiation at AMO amounts to

1 Leiter des Fachbereichs Neue Technologien, Raumfahrt, AEG-Telefunken A 47, Industriestr. 29, 2000 Wedel (Holstein), FRG

1.396 kW/m^2. The spectral composition of solar radiation changes when penetrating the Earth's atmosphere due to dispersion and absorption. The result from this is a shift in the maximum radiation strength toward the red sector, i.e., toward light quanta with low energy. In the case of vertical solar radiation without clouds the power density at sea level and in the vicinity of the equator amounts to some 1.0 kW/m^2 and is described as AM1 (Air Mass 1). But the power density differs considerably according to degree of latitude, the time of year and day, and cloud density on the earth. For example, the maximum rating is reduced to about 0.1 kW/m^2 in case of clouds.

In contrast to the thermodynamic power stations, and thus of considerable advantage, is the characteristic of solar cells to convert direct and indirect radiation into electrical energy with an equally favorable effectiveness. Therefore, when appraising the effectiveness of photovoltaic systems, the integral energy density at the appropriate location of the generator becomes decisive: This varies by more than 100% from 2300 kWh/m^2 annually in desert regions, to 2000 in the south of the United States, 1500 in Southern France, and 1000 in the Federal Republic of Germany.

The extremely exacting quality demands in space flight led to high production costs for photovoltaic systems, ranging from some 500 to 1000 DM/Watt. The sun's lower and also fluctuating degree of energy on the Earth's surface calls for a broadly based development concept to render electricity production costs economically competitive by

  the development of largely automatic production processes for all components of a solar system, from the base material to the power consumers

  the adaptation of all components to the particular location of the units and the consumers by way of an intensive world-wide field study.

The concrete development goal of these programs is aimed at reducing by 1988 the investment expenditure for decentralized generators with a capacity of less than 1 MW to between 2 and 5 DM/W.

The success of developments achieved to date, growing demand, and the future prospects of photovoltaic systems, have led in Europe and the United States to a steep increase in state commitment ranging from development to demonstrations of the various possibilities of application, with the ultimate goal of deriving direct electricity production from solar energy, at a rate of some 3% in Europe and approximately 7% worldwide by the end of the century. This means worldwide installations of a capacity of some 900 GW by the year 2000, which corresponds to a solar cell area of at least 10,000 km^2 and has to be expanded by more than a factor of 2 to account for the requirements under cloudy conditions and during the night.

In the year 2000 electricity consumption worldwide will make up some 11% of primary energy consumption: Given this state of affairs, no spectacular solution of the energy problem can be expected from solar systems; however, the contribution to additional energy is very valuable, since the main exploitation in the Third World, with its future stronger growth rate of energy requirements, will lead to a reduction in overall demand on the international energy market and thus also to the desired easing of demands on fossil fuels.

2 Energy Conversion with Solar Cells

The photovoltaic effect was discovered by Becquerel as early as in 1837, and by 1877 Adams and Day had already recognized the suitability of the semiconductors selenium and copper oxide. The low efficiency, the proportion of reduced electrical energy to the added radiation energy, amounts with these materials to well below 1%, and limits the applications to photoelements for exposure meters in photography. It was not until the development of silicon solar cells with about 6% efficiency in 1954 by Chaplin, Fuller and Pearson that future application potentials opened up for this energy conversion, initially with the dominating role in providing electricity for space satellites. The energy conversion in a silicon solar cell takes place in the sector of the barrier layer between the N-material with a majority of negative electrons and the P-material with a corresponding surplus of positive defect electrons on the charge carrier of the semiconductor (Fig.1). An approximately 0.3 mm thick silicon wafer is doped with phosphorus in order to produce N-conducting material in a boron-doped P-base material. The barrier layer is located at less than 1 µm below the solar cell surface, which is covered with metallic contact fingers as "collector grid." The electrical behavior of the solar cells in a non-lighted condition reflects the dark characteristic curve, which corresponds with the current voltage characteristic curve of an electronic semiconductor diode in rectification mode.

The photovoltaic effect by illuminating a solar cell occurs in three phases:

absorption of light quanta and production of the charge carriers

separation of the charge carriers into P- and N-material

diffusion of the charge carriers to the barrier layers.

The energy of those light quanta which form pairs of electrons/defect electrons in the semiconductor reaches or exceeds a material-specific threshold rating. These charge carriers, formed by light quanta and which are beyond the already available concentration in the thermic balance of the doped semiconductor material, can be separated by the potential gradience within the semiconductor resulting from the differently doped layers of the solar cell. The achieved potential difference between the two sides of the barrier layer corresponds with the open circuit voltage of the solar cell. The separation of the charge carriers is continued by diffusion beyond the barrier layer into the field of direct charge, in the course of which partial

a)

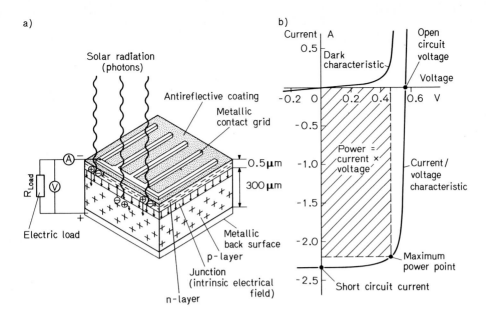

b)

Fig. 1a,b. Basic principle and electrical characteristic of a solar cell.
(a) Schematic description showing a homojunction solar cell. (b) Electrical
characteristic of a 10 × 10 cm multicrystalline silicon cell under AM 1-il-
lumination at 25°C

recombination of these charge carriers leads to a reduced efficiency of the energy conversion. The reason for this are defects
as well as chemical impurities in the semiconductor.

Besides the operation parameters (radiation conditions, temperature), the power of a solar cell is determined by the semiconductor structure (base material, manufacturing process). Therefore, the future success of worldwide development effects depends decisively upon the correct choice of base material and
the appropriate material structure, in which connection the following requirements have to be met:

conversion of solar energy with high efficiency

long life with stable operational behavior

minimum manufacturing costs due to large-scale material availability and energy expenditure

positive environmental rating.

The question of the most suitable base material has also been
discussed among experts, as is frequently the case nowadays with
energy-related problems, in a committed and controversial manner,
many of the arguments have taken on the character of a creed:
With astounding naivety technologies with completely different
potential and stages of development are compared with each other.
The following analysis, therefore, concentrates on the most important base material systems Si (silicon), GaAlAs (galmium-
aluminum-arsenide), and CdS/CU$_2$S (cadmium sulfide/copper sulfide)
(Table 1).

Table 1. Photovoltaic cell efficiencies η (%) of several types of cell design

	Space techn. production	Terrestrial production	Maximum achieved in laboratories	Theoretical maximum
Silicon				
Monocrystalline	14 – 16	11 – 13	19	22
Multicrystalline	–	9 – 11	13	
Amorphous	–	2	6	15
GaAlAs				
Monocrystalline	–	–	22	25 – 28
CdS/Cu$_2$S				
Microcrystalline	–	2 – 3	6 – 9	15

At 30%, silicon is the second most abundant element in the earth's crust after oxygen, and there are practically unlimited supplies of it spread all over the entire earth; at the same time it is environment-compatible and stable. The transformation into pure monocrystal material is very complex and expensive, so that development is concentrated on polycrystalline silicon, which now already represents a convincing alternative. Amorphous silicon absorbs the sun's radiation extraordinarily well, so that, as opposed to the necessary material thicknesses of at least 100 to 200 μm for crystalline solar cells, coating thicknesses of below 1 μm are adequate. As a result of this, material and production costs for amorphous solar cells could be lower provided the hitherto unacceptable low efficiency can be increased yet.

GaAlAs solar cells offer the highest efficiency and can be exposed to very considerable radiation intensities. However, both galmium and arsenic are very expensive, arsenic moreover being a highly poisonous material. The low availability of the base material is evident from the following example: With 1000-fold concentration of solar energy and the processing of 10% of the world's stocks of solar cells, all that could be built up would be a maximum power of 6 GW.

The advantage of low material cost and continuous series production of Cds/Cu$_2$S solar cells presents major and so far insolvable disadvantages: Long-range stability is restricted, due to the sensitivity toward humidity as well as oxygen at 60°C. Over and above this, damage can also be observed by operating voltages over 0.35 V. Finally, resulting from this is a complicated structural element which in addition is difficult to manufacture due to the toxicity of cadmium. The limited availability of cadmium would permit only the buildup of a maximum power of 32 GW.

Taking into account the greatest possible efficiency of solar cell systems, silicon alone as base material in a broad measure meets the demands for large-scale technical application. As long ago as 1974, AEG-TELEFUNKEN for the first time succeeded in proving with a completely new type of material jointly developed with

Wacker Chemitronic, a multicrystalline cast silicon, produced at much lower cost and energy expenditure, that solar cells with an efficiency of over 10% could be manufactured. Of the two modifications in which silicon is present, i.e., amorphous and crystalline, only crystalline silicon is left for short- and medium-term application; nevertheless, in view of its long-term potential, amorphous silicon is being intensively developed.

The energy reflux time as the criterion for energy expenditure is the period in which a solar cell gives up just as much as is required for its production. For solar cells for space use this value amounts to about 40 years. Already in 1978 only approximately 4 years were necessary for terrestrial solar cells, with the subsequent aim of reducing the power expenditure by 1988 to about 1 year, and in the long term this can be further reduced to 3 - 4 months.

3 Technical Design of Photovoltaic Systems

The maximum power of the individual solar cell is comparatively small, and even for the larger 10×10 cm^2 polycrystalline solar cells turned out in serial production to date amounts to only about 1 W (Fig.2). A solar cell generator, therefore, consists of a large number of solar cells, which in turn form smaller units, the modules, for economical production, transport, and assembly. The open circuit voltage of the solar cells is approximately 0,5 V as regards silicon, independently of the cell's size and nearly independently of insolation conditions, so that the desired module voltage of 12 V, 24 V, etc. can be reached solely by series connection of the solar cells. The direct current in short circuit operation, which admittedly is practically independent of the operating temperature, increases almost proportionately to the solar surface area and insolation intensity, so that the desired modular power, in addition to serial connection of the solar cells, is achieved by additional parallel connection. The schematic design of a solar cell generator with all the necessary components into a photovoltaic system is shown in Fig.3. Peak demands as well as electricity consumption in times of deficient solar insolation are overcome by a storage battery which is fitted with a charging and discharging regulator. DC convertors or AC invertors supply the consumer directly. The specific demands of high powered consumers as regards power, voltage, current, and impedance are in general met by way of an electronic power conditioning with a main set conductor configuration, which is characterized by stabilized voltage via a regulator at the generator output.

The chief goal of optimizing photovoltaic systems consists in producing electrical energy in the generator as economically as possible and to insure the rational utilization of electrical energy in subsequent energy storage, power conditioning, and finally, in the consumer. Considerable operating experience is necessary to avoid overdimensioning of the generator for utilization also at small radiation intensities. In future, too, the costs of manufacturing and assembling the complete generator,

Fig. 2. Fig. 2. 10×10 cm^2 multi-crystalline silicon solar cell taken from the series production (by AEG-Telefunken)

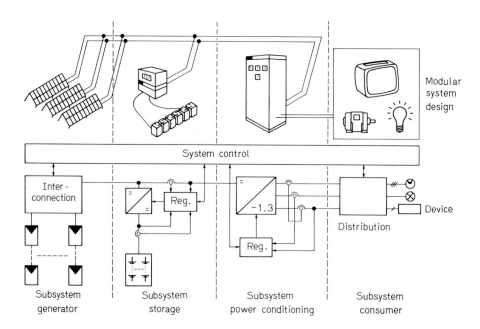

Modular system design

System control

Inter-connection

Reg.

~1.3

Reg.

Distribution

Device

| Subsystem generator | Subsystem storage | Subsystem power conditioning | Subsystem consumer |

Fig. 3. Schematic depiction showing possible components of photovoltaic systems. *Reg.* = Regulator

Table 2. Breakdown of production costs (in %) and commercial prices of generator modules and complete systems

	1980 (Status)	1984/1988 (Development goal)
Base material	7 - 4	5
Si-wafer	11 - 7	15
Solar cell	13 - 9	3
Module	11 - 16	8
Module production	42 - 36	31
Generator (without module)	8 - 9	11
Inverter/Regulator	8 - 9	11
Battery storage	12	9
Remaining devices	8 - 12	14
System production	36 - 42	45
General costs		
(Design, marketing, development, transport, capital costs, etc.)	16 - 28	24

Commercial prices in DM/Wp:	1980	1984	1988
Module	20 - 30	9	3
System	48 - 83	29	8

including the real estate charges, will make up the lion's share of expenditure for photovoltaic systems (see Table 2).

An essential problem of layout lies in the fact that the power of a solar cell generator has to be estimated on the basis of statistical climatic data, so that the insured supply to the consumer demands the use of a storage battery. Optimization of the overall system is therefore reduced to the selection of a suitable concept for the charging operation of the battery, especially, since with the future reduction in costs for the generator the necessary expenditure for energy storage increasingly determines electricity production costs. For each field of application there are two different procedures for charging the storage battery and these differ considerably as regards the effort involved.

The parallel connection of generator and battery requires the least technical effort: The generator must be very precisely proportioned, so that the generator voltage is high enough with the lowest still usable generator power to feed as high a charging current as possible into the battery. This direct charging process should therefore be given preference in very sunny regions with stable weather conditions and minor variations in solar radiation intensity.

With the alternative regulator process to the appropriate maximum utilization of the power provided by the generator, the battery currents and the load currents are supplied via the main regulator. An electronic control system in combination with a large-capacity storage battery retains the excess energy and releases it as required. The operating point of maximum power is the product from the total current and voltage of the generator (see Fig. 1). Despite the greater expenditure and effort for the electronic control regulator, this procedure for photovoltaic systems offers considerable advantages in regions such as Central Europe with large fluctuating radiation intensities.

The application of photovoltaic generators to date has confirmed the particular advantage of the modular construction from a number of similar and simple units, which in addition simplifies future adaptation of the systems to growing power requirements. In principle, one can differentiate between two types of modules:

flat modules with a power density between 80 and 130 W/m^2

concentrator modules with a maximum power density of 150 W/m^2 dependent upon the concentrating factor and efficiency of the optic systems.

In this connection, the demands of the environment have to be insured by an effective life-span of 20 years:

temperature range of $-40°$ to $+90°C$

relative humidity of maximum 90% - 95%

wind loading up to 160 km/h

hailstorm stress up to grains size of approximately 5 cm diameter

Moreover, the module efficiency should not be less than 10% according to the optimizing calculations for complete systems, since the lower solar cell expenditure with lower efficiency is counted by the rising system expense connected with surface area requirements.

The long effective life required for this generator module makes correspondingly high demands on fabrication technologies used for the series and parallel connection and capsulation of the metal contacts and interconnectors of solar cells. According to long-term experience to date, a safety glass system, which embeds the connector solar cell matrix between two glass panes, is the most suitable. Plastic material systems with the encasing of the solar cells in glass fiber reinforced resins at the present time still frequently show failures in damp climates resulting from metallic corrosion and delamination. On the other hand, as against the glass type, the plastic material module results in an additional gain in power of about 7%. The fabrication of economical flat modules demands permanent transformation of every newly developed technology into automatic production for solar cells and modules (Fig.4).

Regardless of the geographic position flat-plate modules are directed rigidly toward the sun and stationarily installed, so that in operation the generator represents a completely static

Fig. 4. First fully-automated production line for generator modules, capacity 2 MW/a (by AEG-Telefunken)

system. The sun-tracking flat generator represents a modification for special demands, particularly in desert, sea, and mountain regions, although the advantage of a 20% to 40% higher energy yield is up against the disadvantage of the higher tracking expenditure (Fig.5). Both generator types with flat modules convert besides the direct radiation also the diffuse solar insolation into electrical energy with the same high efficiency, and thus also exploit a large proportion of diffuse radiation even in southern regions.

Considerable attention was paid to concentrator modules in earlier times because of the still prevalent high costs of solar cells at that time. Optical concentrators with a concentration factor of 20 to 50 increase the energy density of the solar insolation and correspondingly reduce the necessary solar cell surface area. The degree of the module efficiency due to the heating of the solar cells is reduced by way of a cooling medium, the released warmth being utilized as process heating. In this case, the advantages of the static system are set aside and at the same time only the direct insolation portion is used, so that with falling solar cell prices concentrator modules are greatly declining in significance.

Reliable energy storage is of importance for many application forms of photovoltaic systems where the sun's energy availability does not accord with the consumers' requirements. Despite intensive development efforts on the part of battery manufac-

Fig. 5. Tracking solar generator, 3,5 kW with flat-plate modules (by AEG-Telefunken)

turers, to date only lead acid accumulators for power ratings up to several 100 kW are available for an economically long effective life and cycle resistance. In this connection, in particular stationary armor-plated lead acid batteries are suitable because of their high cycle life, although the investment outlays run to some 500 DM/kWh with a maximum life duration of 12 years. In the coming 20 years, too, further development will concentrate on stationary lead acid accumulators and, moreover, special charging and protection electronics with a starter battery is found everywhere in the developing countries. The major investment expenditure and the limited availability of lead as a raw material in the long run call for new types of battery systems, as for example the natrium-sulfur batteries.

4 Decentralized Application of Photovoltaic Systems

The exploitation of solar energy to produce electricity is today, for economic reasons, still mainly geared to applications in southern regions, since the annual radiation density at the equator is, with 2300 kWh/m^2, more than by a factor of 2 higher than in Central Europe. Moreover, measurements of global insolation carried out over several decades proved that in northern regions the year's minimum energy availability in December is only 10% of the year's maximum (May). This ratio increases pro-

Table 3. Examples for the application of photovoltaic power systems

Application	Power range in W peak	Competitive investment costs in DM/W
Measuring stations (land and sea)	10 - 30	100 - 1,000
Small meterological stations		
Ocean buoys		
Alarm systems (fire, water pollution)		
Emergency call boxes		
Sailing boats		
Weather balloons		
Communication systems	10 - 100	50 - 500
Portable transmitter/receiver sets		
Stationary transmitter/receiver sets		
Repeater stations		
Remote-control systems for pipelines		
Educational TV		
Radio relay stations	100 - 1,000	30 - 150
Ground receiver stations		
Small transmitters (radio, TV)		
Signalling installation	10 - 1,000	50 - 200
Road warning lights		
Traffic control systems		
Signal buoys		
Land and sea radio beacons		
Railway control systems		
Lighthouses	from 1,000	(2)5 - 50
Airport lighting systems		
Radar installation		
Independent power supplies	500 - 1,000	50 - 200
Corrosion protection for pipelines and bridges		
Water pumps		
Water desalination		
Cooling		
Single houses		
Village community centers	from 5,000	(2)5 - 50
Decentralized rural electrification		
Military systems		

gressively with approaching the equator, to 30% in Southern Spain and 90% at the equator. Resulting from this, the basic problem is for Central Europe, in addition to the short time storage of electrical energy in storage batteries, also that of securing long-term storage, for example by hydroelectrical systems, or by the integration of photovoltaic plants in the energy supply system. According to this, on a purely theoretical basis, the electricity consumption of the Federal Republic of Germany in 1980 corresponds to a solar cell generator area of about 1% of

the entire surface of the Federal Republic of Germany, while
the surface utilization for buildings, water covered areas, and
waste lands together is already in the range of 10%.

The proven technical, operational, and economic advantages of
utilization to date have enormously reinforced worldwide interest
in photovoltaic systems for decentralized electricity supplies
(Table 3). As opposed to the conventional and other unconventional
energy conversion systems, particularly the following advantages,
from production via installation to operation, should be em-
phasized:

   The largely modular construction of the system's components
   ensures a large degree of flexibility in adapting to growing
   operational demands, making it possible to effect large-scale
   series production and to lower the costs of production.

   The installation of the systems is simple and can also be effected in developing countries on the spot with unskilled
   laborers.

   The photovoltaic system is a static system, so that maintenance requirements can be reduced to a minimum verging on neg-
   ligibility.

   The photovoltaic energy conversion to direct electricity production does not require any cooling medium and can be operated at any location desired.

   The extremely long effective life of over 20 years, by the
   use of high-quality manufacturing technologies, in most cases
   obviates the necessity for substitute investments and concur-
   rent high transport and installation expenditure.

   With a decentralized power supply there are no need for in-
   vesting some 10,000 DM/km for energy distribution and for
   operating costs, since some 15% of the energy produced is
   energy distribution losses.

In developing countries solar cell generators are already now
providing, in many cases for the first time, a solution to the
electricity supply problem in the kW-range and at the same time
represent in the long run the sole alternative for these regions.
With future reduction of the system costs, the application possibilities are increasingly shifting toward greater power ratings, and in this respect preparatory demonstration plants are
being tested: A 60-kW plant to supply electricity to a radar
station in Mount Laguna, California, is already in operation.
Photovoltaic systems in the 300-kW range with flat-plate or
concentration modules are being planned for Europe and Saudi
Arabia. This also accords with the development program on the
way in the Federal Republic of Germany since 1975 with the back-
ing of the Federal Ministry for Research and Technology, of using
on a competitive scale by 1988 solar plants up to 1 MW for economically self-sufficient electricity supplies in villages and
small towns, especially in the developing countries.

Power supplies for electrical appliances for individual house-
holds or for entire village communities are today in the foreground of the economic application of photovoltaic systems in
developing countries, primary attention being paid especially

to operating television sets to improve education in outlying
districts, and to providing illumination, as well as power for
refrigerators. Growing attention is also being paid to solar
air conditioning in the southern regions of the industrial
states. Solar powered traffic facilities in the lower range of
some 10 W are already an economic proposition at the present
time, as for instance buoys for water traffic. Hazard beacons
powered by solar energy are also of great importance, especially
for aviation in outlying areas and where mountains are a major
obstruction to navigation. A significant field of application in
Europe will be also solar powered traffic control. For example,
a solar warning device against the growing problem of driving
vehicles in the wrong direction on motorways, or the first solar-
powered emergency telephone for the "Björn Steiger Rescue Found-
ation", which for nearl 10 years has fitted such telephones on
German trunk roads; also the problem of the kilometer long connections with the grid costing 70,000 DM pro unit: The 0.5 m^2
solar cell generator during the day charges a 24 V battery,
which at night provides power for the flashing light. Thanks to
the solar system, a basic demand is economically solved: Emergency telephones, independent of network connections, can be
installed at points a long distance away from built-up areas.

The largest worldwide utilization today is still solar communication technology with units of several watts for wireless telephones, and going as high as several kilowatts of power for
VHF transmitters. In view of the fact that transmitters for
broadcasting are erected in locations favorable for improving
reception conditions, for example on mountain tops or high
ground, the grid supply is complicated or encounters transport
problems for the working materials if self-sufficient electric
power supply plants are used. Examples are already provided today by television transducers in the Federal Republic of Germany. Following the economical operation of the first plant with
solar electricity supplies, a decision in favor of the solar
solution for another unit was taken. In all cases, therefore,
where data and communication connections have to be maintained
over long distances photovoltaic electrical supply systems are
the obvious solution.

The use of photovoltaic systems for irrigation, for obtaining
drinking water from salt water, as well as for general treatment
of drinking water, meets in particular the fundamental demand
of the population of the developing countries on account of the
frequent lack of infrastructure (Fig.6). For instance, with a
pumping system of 1 kW maximum power and a pump height of some
10 m approximately 30,000 m^3 water can be supplied annually,
and the irrigation of the fields of a medium-sized village is
secured. Furthermore, in many cases the solar generator can be
supplemented by a storage battery, so that the combined use of
radio, television, lighting or other electrical appliances is
possible with reduced pumping power. Drinking water needs are
growing with increasing population and on account of the general
further spreading of civilization and agriculture. For use in
regions with abundant sunshine, the solar powered pressure filter
process for obtaining drinking water was developed, which is
based on the reverse osmosis principle. The unit works with en-

Fig. 6. Rural electrification in Picon/Indonesia, 5,5-kW generator with flat-plate modules (by AEG-Telefunken)

Fig. 7. Water desalination plant in concepción del Oro/Mexico 2,4-kW generator with flat-plate modules (by AEG-Telefunken)

vironment temperature and requires only mechanical energy. With a generator output of maximum 2.4 kW peak a maximum of 175 l of drinking water can be gained from 330 l of groundwater per hour with a maximum of 1% salinity (Fig.7).

The European Economic Community, wishing to demonstrate the advantages of photovoltaic systems, will install pilot units in all member states, with a power range of around 50 kW to a maximum of 300 kW. The largest plant of 300 kW peak for self-sufficient electricity supply of the seaside resort center on the North Sea island of Pellworm, and the neighboring consumers (8 dwelling units, 9 commercial units), will indicate the suitability of a solar system also for northern regions. The load profile of the intended consumers is strongly influenced by the holiday season during the summer-half of the year and around the year's end, although the solar supply of the consumers can to a large extent be harmonized with the global insolation availability of the sun. In addition to the projected self-sufficient supplies with the photovoltaic system, a combined operation with the existing grid supply is to be tested. This pilot project, which up to now has cost some 12 million DM and is based on an annual electricity production worth 150,000 DM, would, with the future solar cell system expenditure of over 2 million DM, already pay for itself within the minimum life duration of over 15 years. The plant is intended to go into operation by the beginning of 1983, and will then represent the world's largest photovoltaic plant with flat modules.

Supplying dwelling houses with solar power, also in northern regions, for example for washing machines and dishwashers, small appliances and electric stoves, as well as to operate heat exchangers for warming the houses, will become increasingly realistic with further reduction in costs for photovoltaic systems. The competent energy authorities in the United States and the European Economic Community (EEC) regard the application of photovoltaic electricity supplies for decentralized dwelling premises as the most important possibility for use on a large scale. In the United States at the present time, for demonstration and testing purposes, single houses and also groups of up to 15 houses are being equipped with photovoltaic generators, which with a power of 4.5 to 7 kW are being completely integrated in the roof structures. The direct current produced by the generator is converted by means of an inverter from 6 to 8 kVA power with an efficiency of some 90% to 95% into alternating current and is connected via a transformer to the house mains. The power of the solar generator and the feeding, as well as drawing from the grid, which in this concept has taken over the task of the energy storage system, are registered by control instruments. In this connection only simple checking and protection facilities are required for regulating the energy flow of the combined generator/grid operation.

A decisive factor for the success of this concept is apart from the expected cost reduction for the solar systems, the compensation at cost price for the surplus solar energy stored in the grid by the utility supply company. From an analysis of the average load patterns in the United States it follows that by the application of photovoltaic single-house supplies, the grid load in the middle of the day can be considerably reduced, so that the extremely expensive current production does not coincide with the peak load. In addition, the American utility supply companies have a major interest in replacing the expensive and ar-

duous grid supplies in outlying areas by decentralized solar
power plants. Quite apart from the different climate in the
United States, the European energy supply boards are likely to
face major problems for the grid operation resulting from the
statistically fluctuating supply of surplus solar energy. The
demonstration projects carried out by the EEC and lasting into
the coming years for the operation of decentralized consumers
in combination with the public supply system will, when the problems are further clarified, provide a relevant contribution also
for Europe.

5 Development Potential of Photovoltaic Systems

The examples for the utilization of solar systems presented so
far are characterized by decentralized operation, which is distinguished from centralized electricity supply by the advantage
that the investment expenses of a contribution system and the
resultant energy losses no longer arise. In the case of solar
power plants with a long effective life there are similarly no
considerable costs for transporting the operating materials,
for maintenance, repairs, and replacement of conventional plants.

These advantages not only ease future introduction to the market,
but enable quite new vistas to be opened up for applications of
solar cell systems; for instance wireless communication connections by solar powered relay stations in arid regions, or electricity supplies for domestic receivers for future satellite
direct television systems in the developing countries.

The medium- and long-range application potential for photovoltaic systems, complementary to and in competition with conventional facilities, depend upon electricity production costs; this
also has to take into account not only the investment expenditure, but also the differences in effective life and the operation costs for maintenance and repair. The fundamental market
analysis is supported by statistic sources and polls and leads
to an estimate of the potential market volume for photovoltaic
systems in the various power sectors, taking into consideration
for the first time also trends in fuel cost (Fig.8). Assuming a
doubling of fuel costs every 7 years, by the year 2000 the present day prevailing difference compared with conventional generators would be reduced by one half.

The market analysis also took into account with photovoltaic
systems the power conditioning and energy storage in order to
achieve a comparative basis for 24-h operation. A representation
with bandwidth is necessary for all systems, since the electricity production costs in each case are dependent upon the power
and the application, location of the plant, as well as the demand of the consumers. This result confirms the medium-range
goal of catching up with and overtaking conventional systems,
dependent upon rising fuel costs, by means of competitive electricity production costs of solar systems below 1 MW; this would
be attained by developing suitable technological processes. Over
and above this, it appears possible by the year 2000 to use
photovoltaic systems also in power ranges up to 100 MW. A precon-

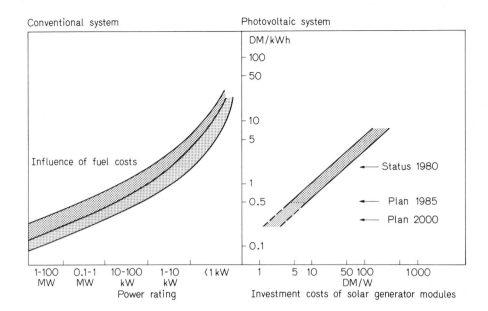

Fig. 8. Photovoltaic energy costs (by AEG-Telefunken)

dition for this, however, are new production technologies for
silicon solar cells, the realization probability of which can
only be judged after achieving the medium-term goals.

The long-term application targets for photovoltaic systems are
the production of electricity on a large scale by competitiveness
with central power stations, both in combination with the existing grid system and also separately, supplying electricity to
a group of small private or major industrial consumers, or to
supply medium and large towns in high solar density regions. The
development and operation of this kind of plant in the multi-MW
range by the method of photovoltaic systems does not pose any
basic technical problems, because the design and operational experience in the multi-kW range can be fully exploited. However,
the necessary profitability cannot as yet be regarded as ensured,
even through potential development possibilities of necessary
technologies, as for instance amorphous silicon solar cells, actually exist. This uncertainty explains the differences in forecasts on applications of central solar power stations, which the
American energy department considers as realizable already by
1990 and not, as indicated in Fig.8, some 10 years later.

Even given a so far assumed successful further development of
photovoltaic systems, the contribution to the long-term solution
of the energy problems in the decades and centuries after 2000
is admittedly extremely valuable, but is of a supplementary and
not a fundamental nature. At this point, model concepts begin
with the goal of providing ideas on a photovoltaic basis both
for central electricity supplies of the conurbations in the
northern hemisphere, and production of fuels for use in road
traffic or in industry as a substitute for fossile fuels.

The concept of a "space power station" was already worked out in the United States at the end of the 1960's and has since then been improved. Today it is regarded from the viewpoint of reliability, as a project which is gigantic but nevertheless capable of execution. The power output of 5 GW at a time into the electricity supply grid calls for a solar cell generator with two surfaces, each of 5.2×4.3 km^2, in geostationary orbit (36,000 km orbit) with a microwave transmission area of 1 km diameter for wireless energy transmission to earth. The reception areas on earth, of 10×13 km^2 size, again convert the microwave energy into direct current or alternating current for feeding into the grid. Estimates for the development, construction and installation of an initial 60 power stations with an overall power output of 300 GW work out at some 1088 thousand million dollars, or 3626 dollars per kW. Similar technical and economic dimensions, which it is true have not yet been examined with the same intensity as "space power stations," also apply to the concept of photovoltaic "hydrogen plantations" in desert regions for hydrogen production, which would be put at the disposal of the industrial states as fuel supplies by way of a gigantic transport system. When ideas of this kind, which reach far into the coming centuries and thus at first sight appear to be futuristic projects, are under discussion, one should not disregard the comparative yardstick, for example of nuclear fusion. Photovoltaic concepts are confronting comparatively minor technical basic problems, and yet a reliable estimate of their probability is for all practical purposes not yet possible at the present time. Nor is this necessary, since further development of the so far defined programs, of necessity, represents also the precondition for the long-range model ideas.

6   Conclusion

This necessarily brief explanation of the photovoltaic concept leads to a future related appraisal of the perspectives for photovoltaic systems, especially since with future cost reductions a strong disproportionately growing market volume will be developed. According to the predictions mentioned in the introduction, by the year 2000 the market will reach an annual production volume in the GW range, assuming that optimistic expectations of further development are fulfilled. For this purpose an industry with billion dollar turnovers has to be built up. Bearing in mind the strong export orientation of photovoltaics, European manufacturers should be able to acquire a strong position.

Wheras at present competitive solar power systems are already being utilized worldwide with a strongly growing trend to higher powers, the future will develop as follows:

Following the conclusion of fundamental developments in low-cost manufacturing processes, introduction to the market for systems up to 1 MW will for the greater part be concluded by 1990.

Parallel with proceeding further with developments and the buildup of large-scale series production, a good chance

exists of pressing on to individual plant sizes of up to
100 MW by the year 2000, although here, too, the decentraliz-
ing aspect is in the foreground.

The contribution of photovoltaics by way of "space power
stations" or "photovoltaic hydrogen production plants," and
thus also to general energy supplies on a major scale, cannot be appraised at present with regard to economic implementation. Therefore, in principle it will be necessary to wait
and see how further development of the earlier mentioned
stages proceed.

The entire future development of photovoltaic systems represents
a chain leading to ever higher power ratings by economic production in successive stages. In present as well as for future
applications it is necessary for the introduction of this new
kind of technology that complete systems, with appropriate train-
ing of future users in installation, operation and maintenance,
and thus also system solutions, be provided by one and the same
person. With the positive development of the photovoltaic method
as assumed, the sociological advantages of decentralized systems
become already apparent, especially for developing countries,
inasmuch as every individual, every village, every town or pocket
state can become indepdent as regards supplies of electrical
energy.

References

Schmidt EF (1975) Unkonventionelle Energiewandler. Elitera, Berlin

Palz W (1978) Solar electricity. Butterworth, London

Proceedings of the Third E.C. Photovoltaic Solar Energy Conference of
Cannes (1980) Reidel Publ Comp, Dordrecht, Holland 1981

Program Assessment Report (1980) Statement of findings/satellite power
systems concept development and evaluation program, Nov 1980. DOE/ER-0085

Exploitation of Wind Energy by Wind Power Plants

S. Helm and E. Hau[1]

In their efforts to exploit renewable sources of energy many countries are increasingly turning to the development of wind power plants.

Unlike other energy sources, the use of wind energy is enjoying a renaissance. After a survey of the past we will try in this article to sketch the present demands which must be met in exploiting wind energy and to describe the latest state of the art.

1 Wind Energy — Its Importance in the Past

1.1 Origin of the Windmill

We find references to the use of wind to drive water scoops and mills in Chinese chronicles dating back to the second millenium before Christ. We do not know, however, what these plants looked like. The first concrete information is given by an Arab chronist writing in the 8th century A.D., who described the use of windmills in Afghanistan. According to the sketches these windmills were increased resistance rotors with vertical axis, similar to those Afghanistanian plants shown in Fig.1, which were still in operation in 1975. Presumably because of their poor efficiency and fixed orientation they were restricted to this extremely windy location and were not copied in the Arab territories or in Europe.

Probably the first mention of a windmill with horizontal axis is to be found in a French deed of gift dated around 1180. The mill was situated in Montmartin en Graine in western France.

From France and the lower Rhine windmills spread in the 13th and 14th centuries along the windy coastal region of northern Europe right up to Russia, and also in the Mediterranean area.

It is noteworthy that even the earliest central European plants showed all the essential characteristics of the typical trestle mill (Fig.2), in particular four vanes of wooden planking or of cloth-covered metal framework and the ability to turn the rotor, with the entire machine room, around a fixed axis into the wind.

---

1 MAN Neue Technologie, Dachauer Str. 667, 8000 München 50, FRG

Fig. 1. Windmill with vertical
axis in Afghanistan

Fig. 2. German Post-Windmill,
since about 1400

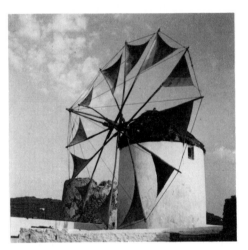

Fig. 3. Windmill with triangu-
lar sails in the Mediterranian
area

Fig. 4. Dutch windmill, 18th
century

In the Mediterranean area, on the other hand, canvas-vaned,
rigid windmills not capable of following the wind (Fig.3) became
common. The classical windmill with horizontal axis was not
adopted by cultures outside Europe. In their perfect form, the
large, high-powered Dutch windmills, developing up to 35 kW
(Fig.4), they were mainly confined to the North Sea and Baltic
areas and to the countries colonized from those territories,
e.g., North America, Australia, and South Africa. In spite of
a wide variety of new developments, such as the American multiblade wind rotor which appeared around 1855, this concept for
high power and severe operating conditions remained unchallenged
until the beginning of the 20th century.

Parallel to the development in Europe, comparatively small machines with vertical axis were in operation not only in Afghanistan but particularly in China, where they were primarily used
for irrigation. They probably had power ratings ranging from a
few watts up to 2 kW.

1.2   Land Drainage in Holland

Where wind conditions were favorable and were consistently exploited, wind energy assumed very great importance. The classic
example is Holland. A large part of that country lies below sea
level, and for centuries it has only been able to be kept dry
by constantly pumping out the water. Right up to the start of
this century the necessary pumping work was performed exclusively
by the aid of wind "mills". Even in the Middle Ages large areas
of land were regained from the sea by building dykes and then
pumping dry the resulting polder. This represented a large-scale
use of wind energy technology, as can be seen from the draining
of the Schermer polder. Between 1631 and 1635 52 windmills
were set up here just for this purpose, pumping out an average
of 1000 m^3 of water per minute at an average pumping head of
3.77 m. From this a net rating of 551 kW can be calculated,
which meant — with the inefficient gearing and pumps then used —
the double or treble power at the rotor shaft. This was for
those days an extraordinary performance, which could not have
been achieved by other means such as the use of men or animals
to drive capstans. Holland, with its large areas lying below
sea level, would not exist as we know it today, nor would its
centuries of comparative prosperity have been possible, without
its wind-driven mills of all kinds.

1.3   The Mill-Builder's Art — Its Contribution to Technical Development

Right up to the beginning of modern times watermills and windmills were the only large machines in widespread use. A large
number of mechanical elements were invented in the course of
mill-building, or tested and perfected to such an extent that
they served as a springboard for the rapidly expanding mechanical engineering industry of the 19th century. Highly stressed
and extremely durable gears, bearings, friction clutches,
hoists, brakes, and control mechanisms were already known in
medieval windmills. In addition to the mill drive, pumps and
grinding, sawing or hammering devices with the appropriate mechanisms were also to be found. In contrast to the watermills,
windmills are characterized particularly by additional installations for engaging and disengaging the very variable energy.
Typical of these are regulators to adapt the vane surfaces via
flaps or to reef the sails, automatic wind-tracking systems
using smaller lateral wind wheels, and powerful servo-assisted
brakes which could stop the mill automatically when actuated.

The development from the first known windmills to the technically refined large Dutch mills was basically a process of constant
evolution, which proceeded fastest in Holland where the windmills

were a vital necessity and were used not only for grinding but
also generally as stationary drive units. From the multiplicity
of applications resulted, almost necessarily, improvements which
were able to assert themselves.

In the beginning, mills were probably copies of plants already
in existence which were erected by local craftsmen. As they increased in size and complexity, the necessary technical knowledge increased too, and, in the case of windmills, to such an
extent that copying was no longer possible. A guild of mill
builders was formed. Despite the large numbers — there were for
a time about 10,000 windmills in Holland and almost twice as many
in the territory covered by the German Empire — the large windmills were never put to large-scale industrial use but always
used as single plants for local craft industries.

In about the middle of the 19th century, the first factory-built
multiblade windmills appeared. This new kind of plant, which
originated in America, supplies in particular a relatively high
torque even when the wind is light. This characteristic made it
ideal for driving small water pumps used to supply the watering
places for the cattle needed everywhere on the American prairies.
At the turn of the century, a respectable wind energy plant industry comprising some 30 companies had been established in the
United States, mass-producing the plants with about 3000 workers.
A total of six million plants was produced by Western Wind Wheels.

Because of the larger, usually constant vane surface, however,
the large versions of this type were in many cases unable to
cope with the wind conditions near costs.

As a result of the unfavorable exploitation of energy and the
high expenditure on gearing as a consequence of low rotational
speeds, the windwheels were unable to compete with the now completely developed windmills in Europe, in spite of the modern
materials of which they were made and of more economic production methods.

2   The Wind as a Source of Energy

2.1   Wind Distribution and Energy Content

The movements of air in the Earth's atmosphere which are per-
ceived as wind are caused by the sun. The wind is thus a secondary form of solar energy. Approximately 2.5% of the solar energy captured by the atomosphere is converted into wind.

If this proportion is calculated in terms of output, the result
is a "wind output" of 4.3×10^{12} kW, far greater than the total
output of all the Earth's power stations.

Both because of global phenomena, such as the rotation of the
earth, and because of local factors, the wind is distributed
unevenly over the earth in terms of both speed and direction.

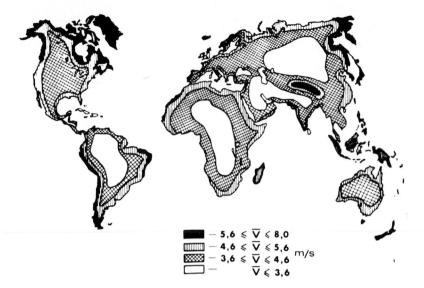

Fig. 5. Windspeed distribution over the earth

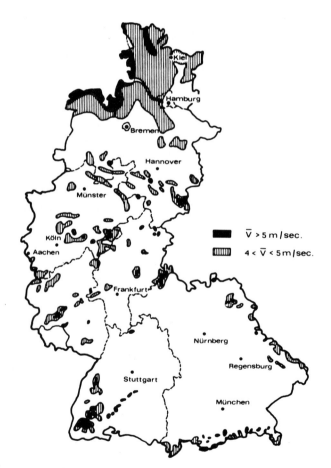

Fig. 6. Windspeed distribution in West Germany

Figure 5 conveys some idea of the global distribution of windspeeds. As is usual in meteorology, the annual averages for a
number of years, measured at a height of 10 m, is given.

Those areas with a mean annual windspeed of 5 m/s and more are
regarded as unreservedly suitable for technical exploitation
of wind energy. The preference for coastal areas and northern
regions is quite clear.

The wind conditions in West Germany are shown in Fig.6. Apart
from the exposed summits inland, the area of interest is limited
to the coastal region in the north.

The distributions of windspeeds provide a preliminary orientation
in matters of the possible use of wind energy plants. In order
to obtain definite statements on the technically usuable energy
potential of a specific place, differentiated data is nevertheless required on the amount and frequency of the wind energy
which statistically is to be expected.

When the so-called cumulative frequency of windspeeds is known
(Fig.7), calculation of the annual exploitation which can be obtained with one wind energy plant is very simple.

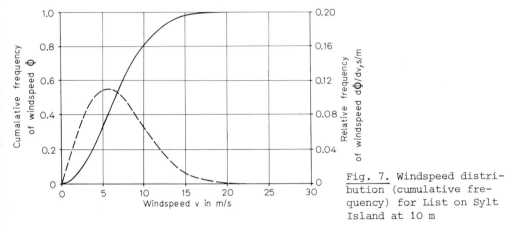

Fig. 7. Windspeed distri-
bution (cumulative fre-
quency) for List on Sylt
Island at 10 m

If the cumulative frequency of windspeeds is considered from
the aspect of the energy content or power contained in an airflow with a cross-sectional area A as follows,

$$P = A \cdot V_w (\frac{\rho}{2} V_w^2) = \frac{\rho}{2} A \cdot V_w^3$$

the distribution shows, in accordance with Fig.8, the energy
content which is to be alotted to the windspeeds according to
their frequency.

It can be seen immediately that, from the point of view of energy exploitation, neither the lower nor the very high windspeeds
are of importance.

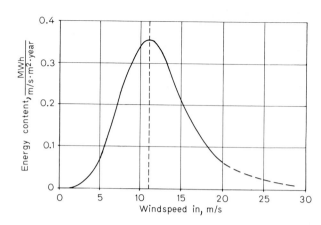

Fig. 8. Energy content of wind per unit area as a function of wind speed (for List on Sylt Island)

The design of a wind energy plant, i.e., the nominal operating point, must therefore be orientated toward windspeeds of some 10 to 12 m/s.

2.2 Characteristic Properties of the Wind

In connection with the use of wind energy in wind energy plants, further properties of the wind are important apart from windspeeds measured over long periods of time.

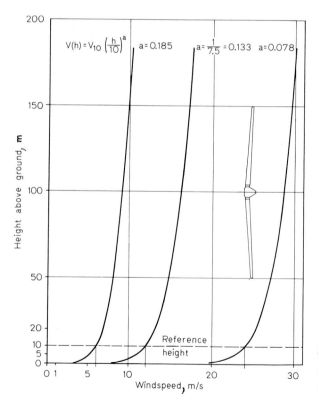

Fig. 9. Increase in windspeed with height (calculated with Hellmann's law of exponents for List on Sylt Island

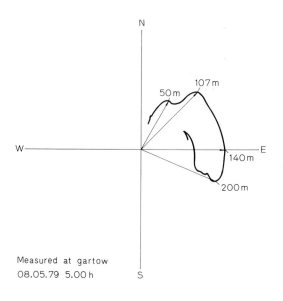

N

107 m

50 m

W———————————————E

140 m

200 m

Measured at gartow
08.05.79 5.00 h   S

Fig. 10. Direction and strength of
wind as a function of height

One of the most important phenomena is the increase of windspeed
with height. Figure 9 shows the increase in the statistically
expected windspeed calculated with Hellmann's law of exponents.
The comparison with the dimensions of the rotor on the GROWIAN
large wind energy plant show that in addition to the fact that
the windspeed has increased at hub height as desired, the speed
of the wind meeting the rotor varies clearly with each revolution. Apart from the gusts which occur at irregular intervals,
this phenomenon is already one cause of the extreme changes in
load to which large wind rotors are subjected. The change in
the direction of the wind as a function of height is another
cause; this is seen in Fig.10.

The brief fluctuations in windspeed — the gusts — are of particular importance when the stresses on a wind-energy plant are
considered (Fig.11). There are still a few areas of uncertainty
here, especially in the case of large wind rotors, as both the
volume and the duration of gusts must be taken into account.

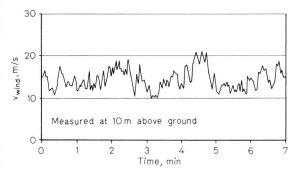

Measured at 10 m above ground

Fig. 11. Windspeed over a period
of minutes

The dimensioning of wind rotors with regard to strength also includes those stresses which are connected with extreme wind speeds
The maximum speeds measured in Europe over the last hundred years
were of the order of 60 m/s ("gust of the century").

The quality of the power and energy delivery of a wind energy
plant is determined to a large extent by the length of time
during which the output remains constant. Apart from brief fluctuations over seconds or minutes, the effects of which can be
cancelled out by the design of the plant or group of plants, the
irregularity of the windspeed over a period of hours is a problem which affects the integration of wind energy into the power
network.

The regulating problems connected with this do not pose a problem for conventional power plants unless the majority of the
supply is derived from wind energy plants. Compared to this, the
annual distribution of windspeeds seems less problematic (Fig.12).

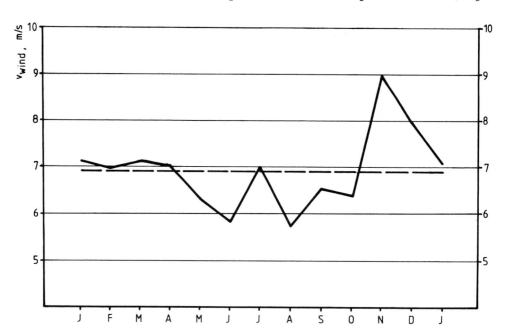

Fig. 12. Average wind speeds in a year. Period of observation: 1969 to 1974;
Site: Busum; Source: German Meteorological Service, June 1978

3   Demands Made on Wind Energy Use Today

3.1   Conditions for its Applicability

The continuing development of windmills up to modern turbines
was interrupted with the rise of steam and other engines. The
reasons for this, apart from the relatively high investment
costs, were the necessarily inconsistent supply of energy and
the difficulty of adjusting to this in an economy based on the

division of labor. Nowadays, energy consumers in practically all fields of application are forced to keep power available at a specific time of their free choice.

The use of wind energy is thus, apart from a few special cases, only possible to the extent to which the demand for freely available output can be met.

A further important pre-condition is that as much of the energy produced as possible is consumed, to ensure that the plant has a sufficient workload.

Potential consumers moreover almost always correspond to specific standards. Wind energy can only be consumed if it is provided at a quality which corresponds to these standards.

In order to fulfil these requirements, it is always necessary that wind energy plants work in some form of arrangement together with energy stores and/or additional sources of energy.

3.2 Fields of Application for Wind Energy

According to the type of consumer, some very different technical solutions with considerable effect on the designs of the wind energy plants are obtained. There are two main groups:

1. Composite systems with another quickly adjustable source of output. The availability and quality of the current is assured in such cases and the workload is partly dependent on the demand. This group includes:

   electricity production where supply is guaranteed when there is no wind by connection to a more powerful grid;

   electricity production in combined operation with an auxiliary current generator which runs in parallel as required or completely takes over supply;

   electricity production in combined operation with a pump store (compressed gas or water);

   heat production; combined operation with an independently regulating heater boiler.

2. Systems which store the required energy. Short-term availability is not necessary in such cases while use is assured to a large extent

   driving pumps for irrigation and drainage;

   pumping takes place when the wind blows; reservoirs and drainage ditches then serve as the store. The demands made on the form of energy supplied and thus on the plants are lower than in the previous cases.

   drive and heating for desalination plants.
   The fresh water tank is the store for the work done.

   pump drive of pump-storage plants with a large reservoir but the wind converter does not feed current directly into the grid.

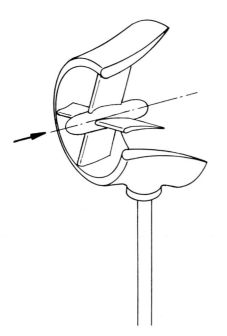

Fig. 17. Ducted turbine

Vortex Tower. A more extensive concentration of the kinetic energy of the wind would be possible with Yen's design for a vortex tower (Fig.18).

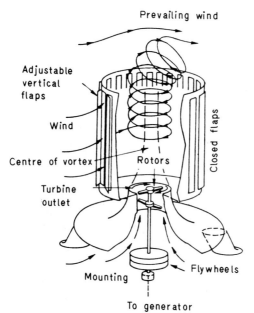

Prevailing wind

Adjustable
vertical
flaps

Wind

Centre of vortex

Closed flaps

Rotors

Turbine
outlet

Mounting

Flywheels

To generator

Fig. 18. Yen's vortex tower

The wind blows at a tangent into a tower through flaps arranged
on the outer casing and creates inside a tornado-like airflow.
Because of the sizeable partial vacuum in the center of the vortex, air is sucked in from the bottom of the tower at high speed,
and this drives a turbine, the diameter of which is about 1/3 of
that of the tower.

So far, this principle has admittedly only been tested in the
wind tunnel, but any attempt to convert the model into a large-
scale design is likely to meet with considerable problems (e.g.,
noise).
Vortex
Vortex Generation by Stationary "Delta Wings". The production of a concentrated vortex is possible by another principle in the form
of a marginal vortex on a stationary "delta wing" (Fig.19). It
is expected that the production of energy by the turbines driven
by the marginal vortices will be ten times that of a free-flow
turbine.

For this system too, only measurements made during wind tunnel
tests are available to date.

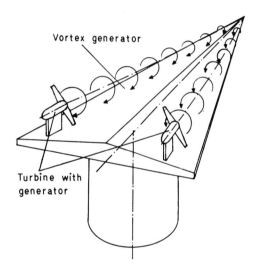

Vortex generator

Turbine with
generator

Fig. 19. Wind turbines with vortex
generator

Convection Power Station. The convection power station is based on
the idea of inducing airflow — as in nature — by heating the air,
i.e., by differences in density. An upward current of air is
created in a high tower situated on a roof covered with solar
panels. This current drives a turbine. This principle is not
really a wind energy plant as such, which uses the naturally
occurring wind, but rather a solar plant which exploits insolation. One advantage of this principle is that it will be possible to use it in those areas which are not accessible to "normal" methods of wind energy exploitation. A test plant with an
output of 100 kW is at present being tested in Spain for the
West German Ministry for Research and Technology (BMFT) (Fig.20).

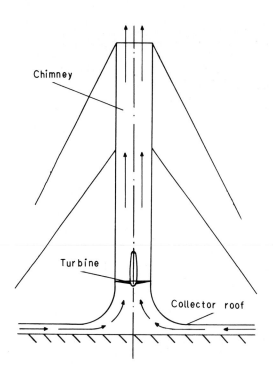

Fig. 20. Convection power station

In the case of the convection plant, as with other special designs which seek to achieve a "pre-concentration" of the wind's
energy, the crucial question is always whether the additional
expenditure on construction can be compensated for by the increase in energy production per unit area covered by the rotor.
Experience to date, although not extensive, does not indicate
that this is possible. The systematic tests being carried out
all over the world at present will, it is hoped, permit more
reliable statements in a few years' time.

5 Planned and Existing Wind Energy Plants (Figs.21-26)

The development of wind energy plants is proceeding with considerable funding in a number of countries. A distinction is
normally made between small plants of nominal output below
100 kW and large plants, which can today produce up to 4 MW. It
is nowadays scarcely still possible to give a complete survey
of the whole range of smaller plants.

Large plants are developed mainly in the USA, Sweden, West
Germany and Denmark, to name only the most important in the
field. These projects are almost exclusively financed by the
state.

In West Germany, the funds allotted to the development of wind
energy technology amount to the princely sum of DM 150 millions.
This amount will be made available by the BMFT over the next
few years for a large number of individual projects for small

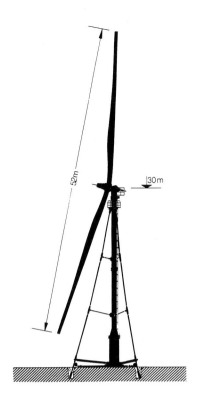

Fig. 21. VOITH WEC 52 at
Stötten. Nominal output
270 kW; Rotor diameter 52 m;
Scheduled for completion in
1981

Fig. 22. MBB wind power plant. Demonstration unit: Nominal output 370 kW;
Rotor diameter 48 m; Scheduled for
completion in 1981

and large plants. These range from research into basic principles via the development of small 10 kW plants to large plants.

GROWIAN is a focal point among the large plants.

GROWIAN, with its nominal output of 3 MW and a rotor diameter
of 100 m, is one of the largest projects in the field of wind
energy exploitation. At the end of 1982, the plant at Kaiser-
Wilhelm-Koog on the lower Elbe should begin trials and feed
electricity into the national grid.

The construction and operation of GROWIAN are being carried out
by a company specially set up for that purpose in which HEW,
SCHLESWAG and RWE are participating.

Apart from the GROWIAN project, another, medium-sized plant
with an output of 270 kW is being developed by Messrs. VOITH,
as is a single-blade rotor designed by Messrs. Messerschmitt
Bölkow-Blohm (MBB). These two test plants should come into
operation in 1981.

Fig. 23. GROWIAN a-
Kaiser-Wilhelm-Koog
(Lower Elbe). Nominal
output 3000 kW; Rotor
diameter 100.4 m;
Scheduled for comple-
tion in 1982

Fig. 24. Danish Nibe wind
energy plant at Aalborg.
Nominal output 630 kW;
Rotor diameter 36 m, 1980

Fig. 25. Swedish WTS-75 "Aeolus" project on Gotland. Nominal output 2500 kW; Rotor diameter 75 m, scheduled for completion in 1982

Fig. 26. MOD-2 at Goldendale, Washington. Nominal output 2500 kW; Rotor diameter 91 m, 1980

Compared with the USA, however, the German wind energy program looks decidedly modest.

A series of prototypes of varying design have been undergoing tests since 1976. These include the largest plant built to date, the MOD-2, with its 91 m rotor diameter and nominal output of 2.5 MW. Since December 1980, three prototypes have been running in trial operation near Goldendale in Washington State. At present, a project is running on the Hawaiian Islands, which will result in the first "wind farm", producing a total of 80 MW from 20 individual plants. This will be the first step toward large-scale technical and commercial use of wind energy plants.

6 Economy and the Contribution of the Wind to Energy Supply

The use of wind energy plants is relevant when the savings on and/or income for the energy produced exceed the building and running costs of the plant. In the past few decades, the price of energy was generally so low that, apart from a few special cases, there was no interest in using wind energy.

Since, therefore, the development, construction and operation of large plants had to compete with this attitude the world over, the empirical values needed for a precise cost-effectiveness analysis are now lacking. Running costs, maintenance costs, availability, etc. can thus only be estimated.

In the wake of the enormous increase in the price of energy over the last few years, the construction of test plants has proceeded apace. The available results can now be used to make the first statements on the expenditure necessary for the construction of large plants; as with other technological products, conclusions are drawn from the expenditure on the prototype and extrapolate these to give the cost of series equipment. Figure 27 shows such a "learning curve", from which the decreasing trend in costs can be seen as a function of quantity, such as can be expected as a result of simplification of design, eradication of errors and rationalisation of production. This shows, compared to the first plant, reductions in construction costs of over 50% for the 1000th plant. The specific manufacturing costs of a prototype in the megawatt range with a load per unit area of approximately 400 W/m^2 (generator output/area swept by rotor) are at present rates between DM 7,000 and DM 10,000 per kW. The construction of wind power stations with several hundred plants would, on this reckoning require DM 3,500 to DM 4,000 per kW of installed capacity.

The costs of electricity production are basically the result of construction costs, capital expenditure, running and maintenance, plant availability and energy production. Estimates made on the basis of comparative values from completed power stations and the above construction costs would suggest electricity production costs of between DM 0.17 and DM 0.21 per kW.

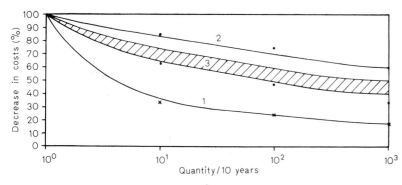

Fig. 27. Decreasing trend in costs of wind energy plants with an output of
1 - 4 MW, as a function of the quantity built in 10 years. Cost of production of first plant: 100%. *1* Data from General Electric; *2* Data from the
crane industry; *3* Assumed decrease in costs for wind energy plants

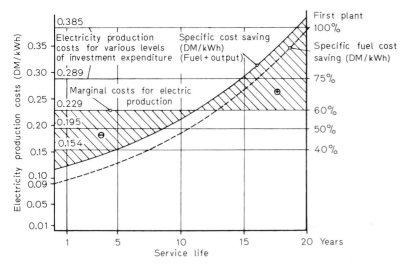

Fig. 28. Electricity production costs for various levels of investment expenditure and specific savings for operation of a wind energy plant as a
function of service life. Period of use 4000 h/a; Spec. fuel costs 0.09 DM/kW;
Fuel price increase 7.5%; Service life 20 years; Interest rate 10%; Annuity
11.8%; Maintenance 1.5%; Tax. insurance 1.0%

If these values were to be compared with the possible cost savings in conventional power stations, and the increase in the
price of primary energy sources taken into account, as is usual
in calculations for power stations (1980 approximately 7.5%),
the result would be economic operation over a specified service
life (Fig.28).

At suitable sites, wind energy plants will pay for themselves,
given the assumed conditions, in approximately 15 years. This
period may seem long, but corresponds in all respects to current
calculations by electricity supply companies .

318

The pre-requisites for the use of wind energy plants in the supply of electricity to consumers of various countries has been investigated using computer models (Jarass 1980). These investigations have also attempted to answer the question of how great the contribution of wind energy to the overall supply of energy in West Germany could be. On the basis of various assumptions which have been made regarding the available area along the German coast and from the problems connected with the integration of the wind energy plants into the conventional electricity supply structure, predictions have been made of the possible share which wind energy could bear.

A share of 5%, given the prevailing economic conditions, is certainly not wishful thinking. The production of electricity with large wind energy plants is in the process of reaching the limit of cost-effectiveness. Since the wind as a source of energy is environmentally harmless, inexhaustible, and independent of supplies from other countries, it will in all probability be used at appropriate sites again in future.

References

1 Eldridge FR (1978) Wind machines. Van Nostrand Reinhold Co, New York
2 Molly JP (1978) Windenergie in Theorie und Praxis. Verlag V.F. Müller, Karlsruhe
3 Windheim R (1981) Nutzung der Windenergie. KfA Sonderdruck
4 Jaraß L (1980) Windenergie. Springer, Berlin, Heidelberg, New York
5 Feustel J (1980) Sonnenenergie und Windenergie, ihr Beitrag an der zukünftigen Bedarfsdeckung. Baustoff - Wärme - Kraft 32 Nr. 9
6 Hau E (1981) Große Windkraftanlagen. MAN - Sonderdruck

Tidal Power Stations

R. Bonnefille[1]

On 26 November 1966, General de Gaulle, President of the French
Republic, opened the Rance Tidal Power Station. Thus an old dream,
that of harnessing tidal energy, first suggested as long ago as
1739 by the engineer Bélidor, started to come true. It was also
the fruit of some 20 years of patient efforts by a French team
that is now scattered.

With an annual output of 500 GWh, the Rance Plant is a small-
scale achievement. But in spite of the fact that it is a prototype, with the operating hazards that this implies, this power
station produces kilowatt hours at a competitive price of about
0.02 US $ (1974) value). This is enough to prove the credibility
of a technique which can make an appreciable contribution to energy production in the privileged countries which have the good
luck to possess suitable sites for using tidal energy.

Tidal energy cannot satisfy the world's thirst for energy; the
most that can be hoped for is to domesticate 400 TWh per year
throughout the world — approximately the equivalent of the annual
energy production of an industrialized country. In fact, tidal
energy is available only in certain privileged sites, where the
range of variation of sea level due to the tide is sufficiently
great — at least more than 4 m (Fig.1) — and where the morphology
of the coast is suitable for the economical building of a tidal
power station. France is favored in this respect, as she possesses
a site capable of providing 25 TWh per year, after some 12 years
of work (Fig.2).

This account of the use of tidal energy will be illustrated by
the example of the Rance Plant, an existing power station, and by
the French project of the Chausey Islands Plant, which has been
under study for nearly 15 years. A few projects elsewhere in the
world will then be considered.

1 The operating Principle of Tidal Power Stations

The Rance Plant is at present the most sophisticated example of
a tidal power station. A dam isolates a sea basin (Fig.3). This
basin must have the maximum possible area for the minimum dam
length; this is the first rule for selecting sites for tidal
power stations. Turbines are installed inside part of the dam,
driven by the water which flows from the sea into the basin when

1 7 rue Anatole France, 78110 Le Vesinet, France

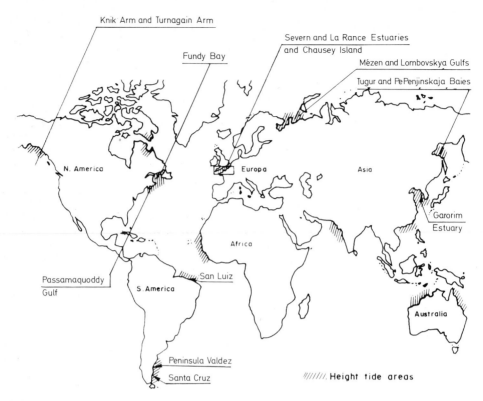

Fig. 1. The potential tidal power plants on the world

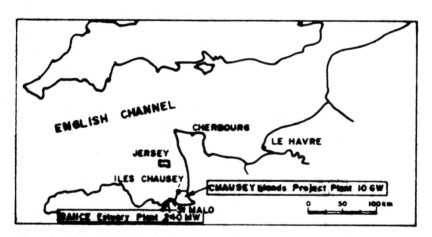

Fig. 2. The French locations for tidal power plants

the tide rises. The basin fills as a result; the turbines stop turning when the difference in level between the sea and the basin is too small (about 1 to 2 m), around high tide. It is then possible either to wait until the sea level has fallen suf-

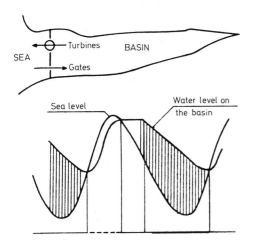

Fig. 3. Operation mode of a tidal power plant with one basin. Hachured areas point out produced energy.
—— Generation; xxxx Standing;
•••• Pumping; ---- Gating

EMPTYING SINGLE-ACTING-CYCLE

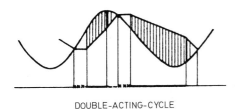

DOUBLE-ACTING-CYCLE

PUMPING WITH OVERFILLING OF THE BASIN

ficiently to use the head between the basin and the sea. Or, and this is an advantage of tidal power stations, if cheap energy is available at the time, sea water can be pumped at low load into the basin, and used a few hours later, when the depth of fall has increased, thus producing two or three times more energy than that used for pumping. In addition, to facilitate the flow of water between the sea and the basin, during the waiting periods at high and low tide, there are large gate sluices in the dam.

This is a flexible system, allowing energy to be obtained at practically any time if the operation of the plant has been pro-

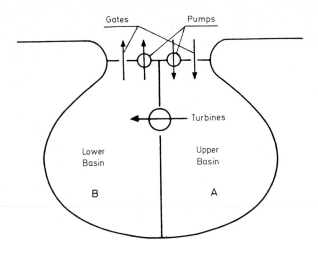

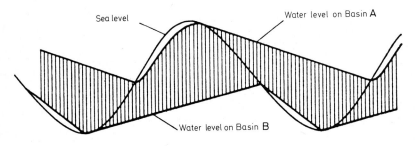

CYCLE WITHOW PUMPING

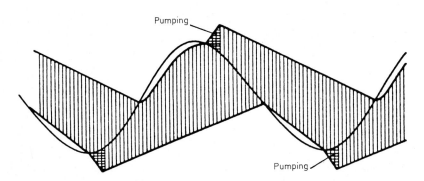

CYCLE WITH PUMPING

Fig. 4. Operation mode of a tidal power plant with two basins

gramed sufficiently far in advance, but it requires relatively complicated reversible turbine units. We may however wonder whether simpler units operated by flow in one direction only (the

323

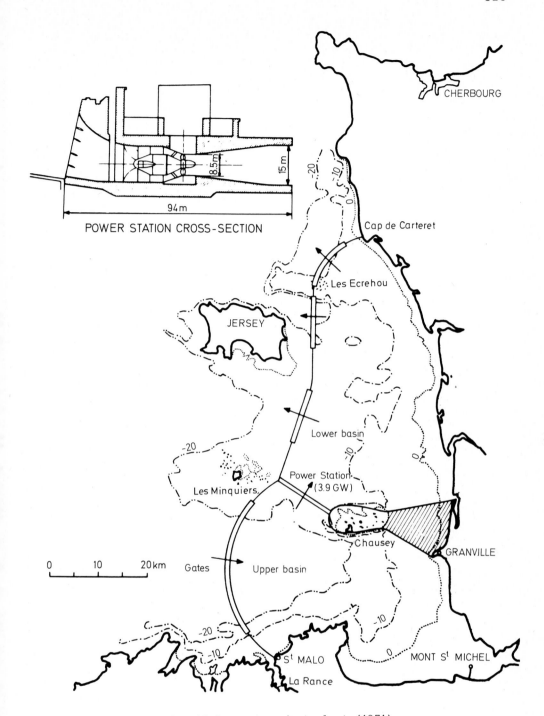

POWER STATION CROSS-SECTION

Fig. 5. Albert Caquot's tidal power project plant (1971)

ebb flow produces more power) would not be more profitable economically in the end.

This type of tidal power station operates only during certain favorable hours (the Rance Plant runs at full output for about 2000 h per year). An attempt has been made to overcome this handicap, by separating the basin in two by a dam containing the turbines (Fig.4). The upper half-basin communicates with the sea when the tide is rising and is filled via gates or even turbines. The lower half-basin empties into the sea when the tide goes out. The turbines use the head of water between the two half-basins, and production can therefore be continuous. The reversible turbine units are conventional, but the civil engineering structures are important. Furthermore, whereas in the conventional Rance-type system the level in the basin varies with the tide, as it did before the power station was built, in the case of a plant using this "Bélidor" cycles, from the name of the inventor, tide ranges are reduced in both half-basins; this obviously has ecological implications which should not be neglected.

Finally, the engineer Albert Caquot has developed a system of operating cycles which uses the energy available during the night from electronuclear units to pump the water so that the tidal power units are ready to operate whenever energy is required, under practically any tidal conditions (Fig.5).

2   The Rance Power Plant

The idea of damming the Rance Estuary to produce tidal power is a very old one (Meynard 1918). Projects were drawn up in 1897, 1902, 1907, and 1920. The studies which led to the opening of the present plant in 1966 started in 1943, under Robert Gibrat.

The Rance Power Station has a single basin, in an estuary 20 km long separated from the sea by a dam 700 m long, equipped with turbines and sluices (Fig.6). The originality of the Rance Power Plant is that it can be adapted to the optimum type of operation. A single-acting-cycle can be used, preferably by harnessing the flow from the estuary to the sea, when the level of the sea is sufficiently lower than that of the estuary: the disadvantage of this is discontinuous production. A double-acting-cycle is also possible, running the turbines when the basin is filled by the flood tide and then, when it is emptied, by the ebb; production is then more continuous, but there is less power because the heads are lower than with the single cycle. Finally, around high tide, when cheap energy is available this can be used to "over-fill" the basin; this energy is then recovered a few hours later with an efficiency of more than 2.

Operating a tidal power station thus means finding the optimum sequence of periods of running the turbines, gating, standing and pumping, taking the following factors into account:

the tide forecast (maximum range 13 m at the Rance Estuary);

the number of units available in the plant (24 × 10 MW units at the Rance);

the number of sluice gates available (six at the Rance Plant);

energy costs;

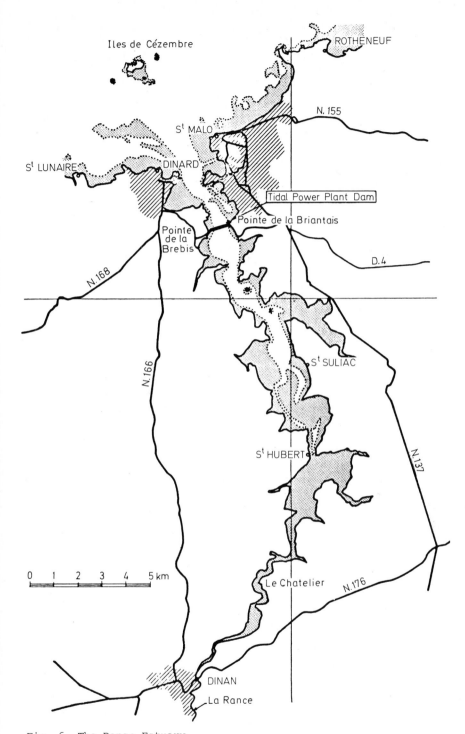

Fig. 6. The Rance Estuary

technical limitations on energy transfer due either to faults
in the interconnecting networks or to the unavailability of
pumping power at certain times;

limitations designed to preserve the environment, for example
a ban on raising or lowering the level in the basin too much
and too often compared with the natural movement before the
power station was built;

safety limitations — sudden changes in the level of the basin
may have to be avoided as they could disturb shipping and make
the neighboring beaches dangerous.

An optimum operating schedule is therefore drawn up in advance,
based on data that can be forecast (tides and estimated energy
costs). This schedule is adapted as unavailability of equipment
and fluctuations in the energy market become known. Finally, the
program is corrected during operation according to meteorological
factors, in particular tide levels that are not as forecast, and
incidents in the environment of the power station and in the
electricity network.

From a study of the operation of the Rance Power Plant (Bonnefille
1976), the results can be summarized as follows:

the power station cost 170 million US dollars (1974 value);

it produces 0.5 TWh per year;

this output is distributed over approximately 2000 h;

the cost per KWh was 0.02 US $ in 1974;

the equipment is highly reliable — unavailability of the units
is less than 5%.

Furthermore, the 20 years of efforts, persuasion and studies by
the team which built the Rance Plant also led to:

the invention of a doctrine for the use of tidal energy: the
theory of cycles and programing of the operation of the plant
outlined above (Gibrat 1953);

the development of bubble units;

the updating of a hydroelectricity technology using sea water
(La Houille Blanche 1973).

3   The Chausey Islands Project

While the Rance Plant was being built, Electricité de France was
studying a tidal power project on a quite different scale, 50
times bigger than the Rance one. This involves separating a 600
to 700 km^2 basin from the sea by more than 30 km of breakwaters
in which a power of 10,000 MW would be installed, producing 25
TWh a year. The project was first outlined in 1942, but when the
execution stage arrived, in 1965, the work was halted. The history of these 20 years of studies is marked principally by the
development of bubble units and by the estimation of the effect
of the plant on tidal currents in the Channel (Bonnefille and
Chabert d'Hieres 1967), taking into account the Coriolis force
due to the earth's rotation.

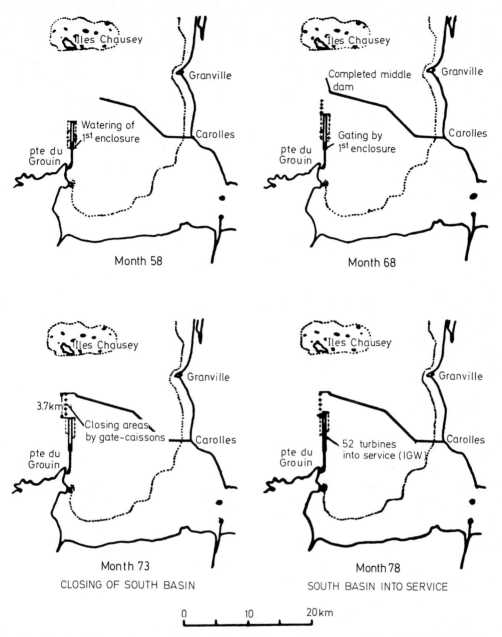

Fig. 7. Building steps of the GTM's 1957 project years 1 to 6.5. ++++ Gate
caissons; ---- station caissons; ▨ gates; ▭ power station; —— dam

The project was updated in 1975. The power station would have
three hundred 40 MW units, it would be built in 12 years, production starting in the 6th year (Figs.7 and 8). The total cost
was estimated as 5000 million US dollars in 1974.

There are two possible construction techniques:

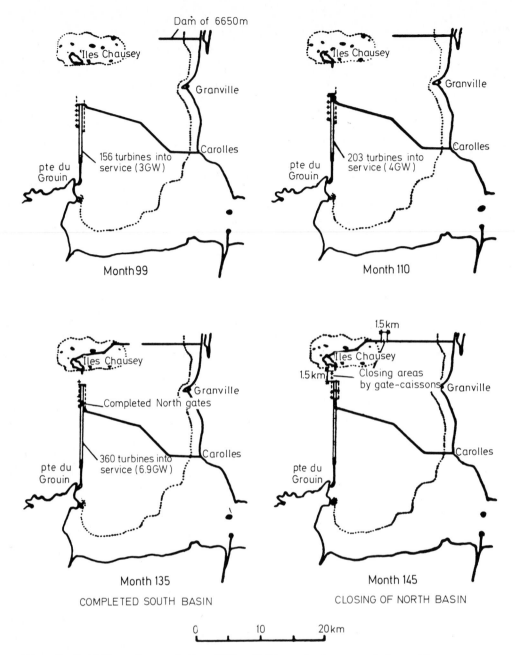

Fig. 8. Building steps of the GTM's 1957 project years 6,5 to 12. ++++ Gate caissons; ---- station caissons; ▨▨▨ gates; ☐☐ power station; —— dam

building a dam plant like that at the Rance, in pumped-out enclosures;

or building caissons on land, sized 100 m × 100 m × 30 m, and submerging them.

Two figures show the vast scale of the project — 20,00 tonnes of concrete per day would be poured for 6 years, and each week a new 40 MW unit would be connected to the network.

But two obstacles remain to be overcome. One of them is technical, and involves choosing between two operating systems:

Either the plant will use single or double-acting-cycles, as at the Rance Power Plant, which will have little effect on the environment, as normal tidal movements will be preserved in the basin, but this will result in discontinuous production dependent on the tides.

Or it will use the Bélidor cycle, i.e., the turbines will be installed between two basins of different levels. In this case the same power output, more easily programed, could be obtained with a lower installed power (nearly half). But in the upper basin the shore would never be uncovered, and in the lower basin the tide would always be low!

The other obstacle is economic. The plant could supply the French network with 25 TWh in 1995. If we extrapolate the electricity consumption, Electricité de France should then be producing 300 to 400 TWh per year. In 1995, therefore, the tidal power station would supply only the annual increase in demand. As there is only one tidal power site of this scale in France, although it is competitive from the point of view of price per kWh produced, we may wonder whether the investment involved would not be better spent in developing a less limited source of energy. This handicap would disappear if future electricity consumption became much more reasonable.

4   Other Tidal Power Station Projects

The criterion for selecting a tidal power site is the relationship between the annual energy production that can be expected, and the size of the structures, which can be represented by the length of the dams separating the basins from the sea. Very roughly speaking, we may estimate that 10% to 20% of the "natural energy" of the site can be converted into electric power. The natural energy is defined by the following equation:

$$E = 2 \, S \, A^2$$

in which E is in kWh per year, S is the area of the basin in m^2, and A is the average tide range in meters. The Rance Estuary has a natural energy of 3 TWh and the Tidal Power Plant produces 0.5 TWh per year: the Chausey Islands Project, with a natural energy of 100 TWh, should produce 25 TWh per year. For these two plants the natural energy in relation to the length of the dams is approximately 4 TWh per kilometer.

Among equally and even more interesting projects that have been studied (Fig.1) we may mention:

the Severn Estuary in Great Britain;

the Fundy Bay in Canada;

the Passamaquoddy Bay in the United States of America;

the Valdez Peninsula in Argentina;

the Tugur and Penjinskaja Gulfs (Okhotsk Sea, 130 TWh per year) and the Gulf of Mezen in the White Sea (40 TWh per year) in the U.S.S.R.;

the Garorim Estuary in South Korea.

The project which seems to have been most studied and is there-fore most likely to be carried out in the near future is the Canadian project for the Fundy Bay between New Brunswick and Nova Scotia. Two sites have been selected, with the following charac-teristics (Fig.9):

Cobequid Bay: 3800 MW, 12.6 GWh per year
               to be built in 11 years
               cost 9290 million US dollars
               saving 1.3 million tonnes of coal per year

Cumberland basin: 1085 MW, 3.4 GWh per year
               to be built in 7 years
               cost 3120 million US dollars
               saving 380,000 tonnes of coal per year

The profitability thresholds of the projects are about 30 to 35 years.

The conclusions of the Canadian studies are as follows:

a tidal power station can replace some of the energy from fossil fuels, but does not make it possible to eliminate or reduce production of electricity from nuclear power stations;

the harnessing of tidal energy is economically profitable, but because of the very heavy investment required the governments concerned must help to provide the necessary capital;

the pursuit of the studies is justified, going on to the stage of detailed technical design and social, economic, and ecolo-gical studies.

The British Severn Project, in its present stage, involves dam-ming the estuary to make a large 450 km^2 basin to serve as the upper basin, with a small 50 km^2 lower basin. Turbines could be run by the flow either from the large basin into the sea or into the small basin, or from the sea into the small basin (Fig.10). Water level could be raised in the upper basin or lowered in the lower basin by pumping. The installed power in the Severn Estuary Plant would be 5000 MW. The plant would be run by pumping at night, and resupplying a guaranteed power of 3500 MW for 12 h or more 1 day out of 16, or 4 working days during the three winter months. The price of the equipment is estimated as about 1200 million US $. The energy balance, after allowing for pumping during the night, would be about 1.2 TWh per year (42 TWh pro-duced during the day and 40.8 TWh consumed during the night). One realizes also to what extent this is a plant for storing en-ergy by pumping, with the contribution from the tide adding 2.5%. A summary of the studies has been published (Shaw 1980).

Argentina is another country favored by tidal energy. It possesses one of the greatest tide ranges in the world at Santa Cruz (12.1 m)

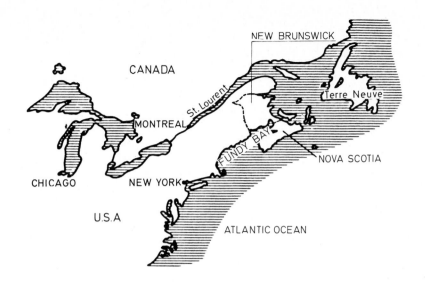

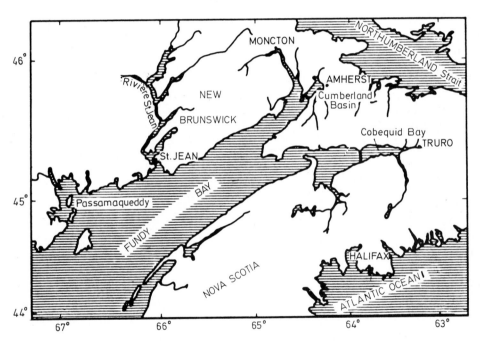

Fig. 9. The tidal power site area of Fundy Bay

where a 6000 MW project has been drawn up. There is an excep-
tional site at the Valdez Peninsula between the Gulf of San José
and the Nuevo Gulf. On either side of an isthmus 6 km wide there
is an almost permanent difference in sea level of some 3 to 5 m
because of a time lag of 4 to 6 h between the high and low tides.
A 600 MW project producing 1.5 to 2.5 TWh per year could be built
there. The two Argentinian projects are economically attractive,
but they have the following arguments against them:

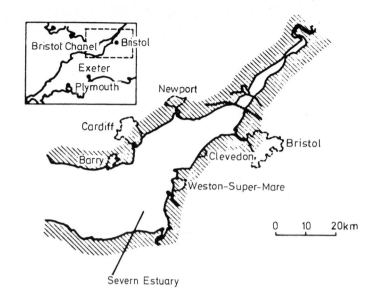

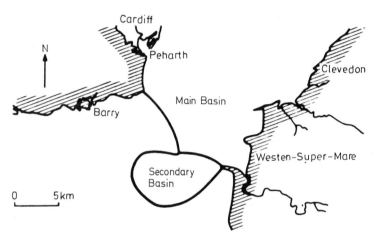

Fig. 10. The tidal power site areas of Severn Estuary

they are several thousand kilometers from Buenos Aires, where energy consumption is concentrated;

Argentina still possesses a hydroelectric potential which can be developed more cheaply;

finally, in the case of the Valdez Peninsula, studies have not yet gone far enough for the feasibility of the project to be asserted. In particular, would the tidal time lag be changed in such a way as to reduce the effective head when the plant is in operation? This is to be feared; the studies for the Chausey Islands Power Station, for example, have shown that the fact of building and running the plant would reduce the tide range by nearly 1 m.

Another South American country, Brazil, is interested in tidal
energy. Along the coasts of the north-eastern provinces of Brazil
the tide range exceeds 8 m, and the coast is indented by a number
of estuaries. There is considerable mining wealth to be processed,
and there is no conventional energy nearby. All this explains why
Brazilian electricity companies are interested in these sites. A
rough project for a small experimental plant near San Luis in the
State of Maragnan is being drawn up.

The characteristics of the Russian projects are as follows
(Bernstein 1979):

Site	Average tide range (m)	Length of dam (km)	Area of basin (km^2)	Installed power (GW)	Annual production (TWh)
Gulf of Mezen	9	86	2215	10	32
Penjinskaja Bay 1	7	72	19000	100	300
Penjinskaja Bay 2	7	32	6320	35	100
Tugur Bay	8	36	1800	9	25

The large tide ranges in the China Sea (6 to 9 km at spring tide)
have led South Korea to choose a site in the Garorim Estuary
where a 2 km dam could shut off a 120 km^2 basin at high tide, and
where natural energy is 3400 GWh per year. A feasibility study
for this project is being made in 1981, and the installed power
would be about 400 MW (KECO 1978).

5 Conclusion

It is by no means obvious that, in the case of a plant with a
low output compared with that of the network, it is useful to
build a tidal power plant as sophisticated as the Rance Station,
i.e., capable of every type of operation — direct and reverse
turbines and pumping, etc. The Rance Plant was considered as a
prototype, and as such it has borne fruit. But a plant running
on the single-acting-cycle (as the Rance Plant has done since
1975 for reasons of expediency) would produce about 5% to 10%
less power, and would be equally competitive. Furthermore, the
disadvantages of non-continuous production from a tidal power
station would be less important than a quarter of a century ago.
Finally, if energy becomes scarce, we should not rule out the
possibility that consumers, and hence industry, will, as in the
past, submit themselves to the natural rhythm of energy source.

The majority of projects are for large-scale sites. In the industrialized countries, tidal power projects come up against
the development of nuclear power (France) or difficulties in
financing building (Canada). Elsewhere they are far from centers
of consumption and their only chance is that local heavy industry
will develop, complying with the known-in-advance variations of
tidal energy production.

To sum up, thanks to the Rance Tidal Power Plant and the recent development of methods and equipment for offshore work (a technological spin-off from exploration for and transport of oil) building a tidal power station is no longer an adventure. Tidal energy will not quench the world's thirst for energy, but it is not negligible, and it remains a standby if critical energy situations arise in the next half-century.

References
═══════

Bernstein L (1979) Il est temps de construire des usines marémotrices. Energo-export 8:15-19

Bonnefille R (1976) Les réalisations d'Electricité de France concernant l'énergie marémotrice. La Houille Blanche 2:87-149

Bonnefille R, Chabert d'Hieres G (1967) Etude d'un modèle tournant de mer littorale, application à l'usine marémotrice des Iles Chausey. La Houille Blanche 6:651-658

Gibrat R (1953) L'énergie des marées, Bull Soc Electr 7 III 29:283-332

La Houille Blanche (1973) Six ans d'exploitation de l'usine marémotrice de la Rance. 2-3

KECO (1978) Korea tidal power study. Korea electric company, Phase I. August 1978

Meynard E (1918) Etude sur l'utilisation des marées pour la production de la force motrice. Rev Gén Electr Nov/Déc: pp 653-658, 607-715, 749-762, 793-802, 823-843, 865-877, 903-914, 947-959, 977-1007

Shaw T (1980) An environmental appraisal of tidal power stations. Pitman Books Ltd, London

Geothermal Energy

O. Kappelmeyer[1]

1 Introduction

Geothermal energy is the natural heat of the earth. The total amount of heat stored within our planet at temperatures above the climatic mean annual temperatures at the surface is of the order of 10^{31} J. This quantity of energy is inexhaustible by any technical use (the present technical energy consumption of the world is of the order of 8 TWa $= 2 \times 10^{20}$ J).

Geothermal energy has been used since the early times of history for therapeutic baths (balneology) and space heating. Since the beginning of this century applications were extended to the generation of electric energy (in Larderello, Italy, since 1913) and municipal, industrial, and agricultural heating. Geothermal energy developments for power generation have since occurred in a number of countries including Italy, the United States, New Zealand, Philippines, Japan, Mexico, El Salvador, Indonesia, Iceland, and Turkey, reaching a total world electric capacity of about 2000 MW in 1981. All occurrences of terrestrial heat, which can be used for electric power production with available and proven technologies, are located within the volcanically active regions of the earth. Figure 1 shows the location of geothermal power stations and the zones within which there is a potential for the economic exploitation of terrestrial heat for electric power production. These zones are called the major geothermal belts of the world.

The incentive for developing geothermal energy was small when hydrocarbons were inexpensive and presumed in abundant supply. Furthermore it has to be taken into consideration from the beginning of any geothermal exploration activity that heat cannot be transported economically over long distances. The energy content of a unit of volume in oil, gas, coal, or nuclear fuel is much greater than that of the same volume in hot water or steam. Therefore terrestrial heat can be used only close to its production site economically for space heating or power production. A further economic burden arises from the content of minerals, acids and non-condensable gases in the geothermal fluids from the underground. These components might cause corrosion and scaling in the production holes, pipes, and heat exchangers.

Similarly to oil and gas exploration, the exploration for geothermal resources involves an investment risk. The combined dis-

1 Bundesanstalt für Geowissenschaften und Rohstoffe, Postfach 51 01 53, 3000 Hannover, FRG

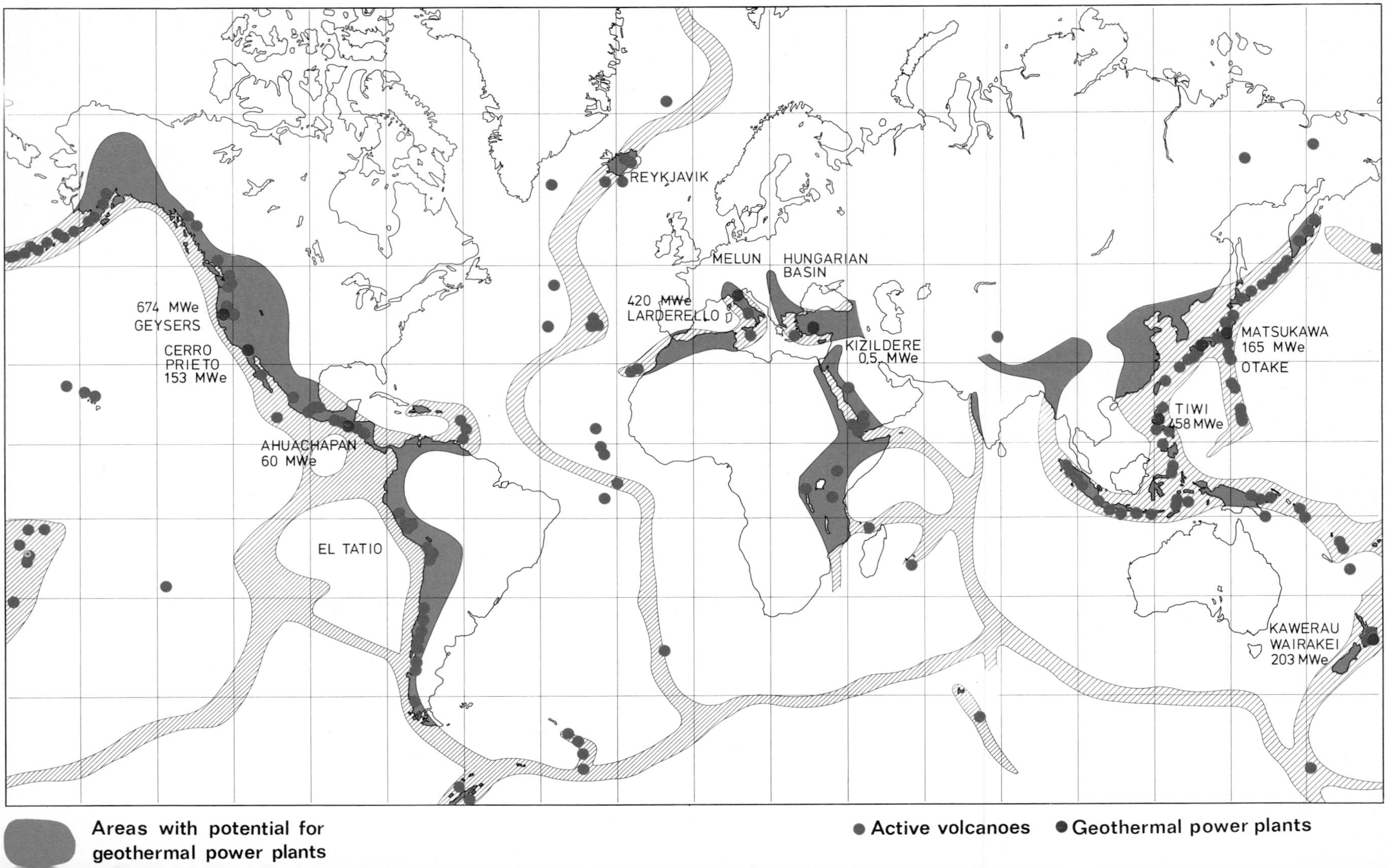

MATSUKAWA
165 MWe OTAKE

KAWERAU
WAIRAKEI
203 MWe

TIWI
58 MWe

KIZILDERE
0.5 MWe

MELUN HUNGARIAN
BASIN

REYKJAVIK

420 MWe
LARDERELLO D

674 MWe
GEYSERS

CERRO
PRIETO
153 MWe

AHUACHAPAN
60 MWe

EL TATIO

● Active volcanoes   ● Geothermal power plants

Areas with potential for
geothermal power plants

advantages of the necessity of risk capital and the high cost
for transportation of heat are a further reason for the long
neglect of exploration of geothermal resources. Recently worldwide interest in geothermal exploration and the development of
new technologies for extraction of terrestrial heat under less
favorable natural conditions are increasing fast. These activities include the exploration of natural steam within the geothermal belts of the world, the exploitation of deep-situated
water-containing layers for space heating, and the development
of technologies for the extraction of terrestrial heat from low-
permeable, dry, hot rock sections. Mainly the extraction of heat
from hot, dry rock would significantly increase the contribution
of geothermal energy to the world energy supply.

The contribution of geothermal energy to the total technical energy production of the world will be small especially in the
highly industrialized countries in the coming decade. However,
in developing countries, which are situated within the geothermal
belts of the world, geothermal resources are an important, and
sometimes the most important, indigenous source of energy. In El
Salvador geothermal power already supplies one third of the electric consumption and natural conditions are promising a far
greater potential. The Philippines is one of the leading nations
in developing geothermal power production. The present electric
capacity is of the order of 400 MW; plans for the next years
call for 2000 MW. Fast developments can be expected in other
countries too (see Table 1).

The cost for geothermal power stations is of the same order as
for hydropower stations and also the production cost for electricity is about equal if hydropower stations can be operated
at the same high annual production-time-load factor as is usual
for geothermal power. A great advantage of geothermal power stations is that they are most economical in the range from 10 to

Table 1. Electric power production from geothermal energy

	1981	Planed for 1985
U.S.A.	1000	2000
Italy	500	600
New Zealand	250	
Japan	250	1000
Mexico	200	400
Philippines	400	2000
El Salvador	60	100
Iceland	65	
U.S.S.R.	5	25
Turkey	1	15
Nicaragua		200
Indonesia		200
Chile		100

Fig. 1. Geothermal belts of the world — existing geothermal power stations

100 MW units. In many developing countries this is also the range of increasing electric energy demand in regions which can be supplied from a power center.

2 Classification of Geothermal Resources

Temperatures of the earth's crust increase with depth. The normal geothermal gradient is between 25° and 30°C/km depth, which means that a hole drilled to 1000 m depth reaches rocks whose natural temperatures are 25° to 30°C above the mean annual temperature at the surface. The mean annual surface temperature is determined by the climate (respectively the sun). The influence of terrestrial heat (the heat flux from the subsurface) upon the temperatures at the earth's surface can be neglected. As a consequence of the increase of temperatures with depth, heat is conducted from depth to the surface. The world average of this terrestrial heat flow is 63×10^{-3} J m^{-2} s^{-1} or 10^{21} J per year for the whole earth. This energy is supplied by the decay of the long-lived radioactive isotopes uranium ($[^{238}U]$; $[^{235}U]$), thorium ($[^{232}Th]$) and potassium ($[^{40}K]$) in the subsurface. The temperature increase with depth (the geothermal gradient) can be derived from the terrestrial heat flow q and the thermal conductivity of the rocks k: geothermal gradient (grad$\vartheta = q/k$). For constant terrestrial heat flow q the increase of subsurface temperature is high in materials with low thermal conductivity. In comparison with other materials, rock has only a very low thermal conductivity. For instance, the thermal conductivity of metals is higher than that of rocks by the factor of 100 to 1000. As a consequence of this low heat conductivity of rocks the temperature rise is relatively great with depth, although the terrestrial heat flow connected with this is extremely low. The range of variation of thermal conductivity occurring with various rock types is shown in Fig.2. Due to the low values for thermal conductivities of rocks it becomes possible that high temperatures occur in depths that can be reached by technical means (boreholes: maximum depth 10 km; shafts: maximum depth some km). On the other hand, the extraction of terrestrial heat from dry, impermeable rocks is inefficient, because rocks behave rather like thermal insulators. The extraction of technologically important amounts of heat from rocks by means of heat conduction — and this is the only possibility for the extraction of heat from impermeable, dry rocks — requires very large contact interfaces between the rock complex that is to be cooled down and the heat transfer medium. Much easier is the heat extraction in the course of production of water or steam from porous and permeable layers in the subsurface. If such aquifers are situated in zones with high temperatures — e.g., in the vicinity of recent subvolcanic activities or in great depth — high yields can be obtained because of the large natural surfaces between water and rocks and the usually long residence time of the thermal waters along these contact surfaces. The degree of concentration of thermal energy in a porous, permeable geothermal reservoir is obvious if the heat content of crustal rocks in the upper 10 km, 85 kJ/kg, is compared to the enthalpy of saturated steam (at 236°C and 3.2 MPa), 2790 kJ/kg. This demonstrates again the advantage of natural occurrences of hot geothermal fluids against

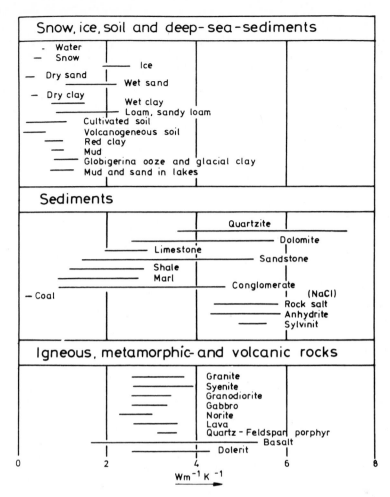

Fig. 2. Thermal conductivities of various rock types

heat in dry rocks. The concentration of geothermal fluids in the underground requires hot layers with high porosity (storage co-efficient), high permeability (hydraulic conductivity), and a large amount of recharge water into the reservoir.

The economic value of geothermal heat is dependent on the temperature. Figures 3 and 4 show the most important possibilities for the utilization of heat at different temperatures. A geothermal occurrence has an economic potential only if great amounts of heat at relatively high temperatures (at least 65°C for heating purposes and 150°C for generating electricity) are available. The zone of high temperature must lie at a depth that is within reach of boreholes — at the present stage of drilling technology not deeper than 10 km — and the energy flux per well must be great if an adequate return on the capital used is to be obtained. Only in areas heated by volcanic activities are temperatures of several hundred degrees Celsius found within a few kilometers of

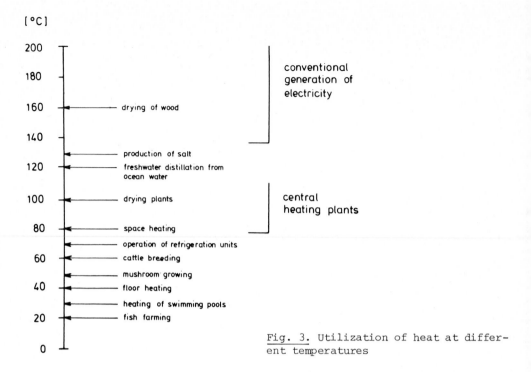

Fig. 3. Utilization of heat at different temperatures

the surface. Therefore, the geothermal fields that have been developed so far for the purpose of generating electricity are found only in the active volcanic zones of the earth.

The various geothermal occurrences can be classified according to their thermodynamic and hydrological properties which result from their geological environments. The following broad categories are usually used.

I *Convective geothermal systems*

a) Hydrothermal systems in high-porosity/permeability environments mostly related to shallow, young silicic intrusions. Technologies for the exploitation of these resources and their use for electric power production are available.

b) Hot magmatic systems — the huge heat content at high temperatures of magma intrusions. No technologies exist for their use.

c) Circulation systems in low-porosity/fracture-permeability environments in areas of high to normal regional heat flow. Mostly used for heating purposes.

II *Conductive geothermal systems*

a) Low-temperature/low-enthalpy aquifers in high-porosity/permeability sedimentary sequences in regions of normal to slightly elevated terrestrial heat flow — these resources can be used for space heating or process heating only.

341

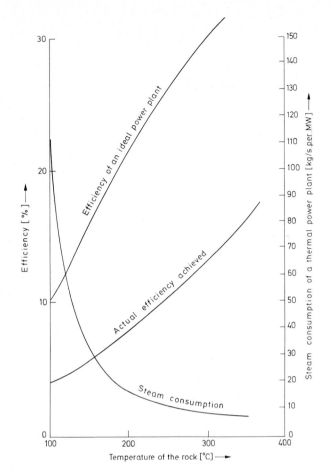

Fig. 4. Efficiency of geothermal power plants

b) Hot Dry Rock in a high-temperature/low-permeability environment — the technologies for heat extraction are being investigated.

c) Geopressured reservoirs, containing pore fluids and natural
gas at very high pressure, up to the pressure of the overburden (= lithostatic pressure); their energy source is heat,
hydrocarbons, and kinetic energy. The energy potential of geopressured systems is great; their use presumes the development
of new technologies.

Figure 5 shows typical temperature graphs for various geothermal
systems.

The geological and geothermal structure of a hydrothermal, high-
enthalpy deposit whose heat content is sufficient for the generation of electricity is shown in Fig.6. For the formation of geothermal, high-enthalpy deposits with reservoir temperatures above
150°C, a huge, natural heat source is required within a depth of
several thousand meters. Therefore, the regions with a potential
for the formation of geothermal deposits with high enthalpy are
limited to zones with recent (Cenozoic) volcanic activity.

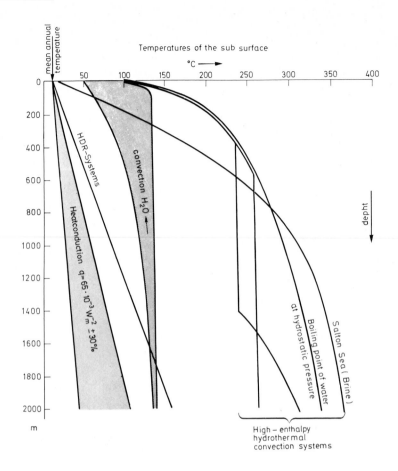

Fig. 5. Typical temperature graphs for various geothermal systems

A further hydrogeological requirement for the formation of a geothermal reservoir results from the tremendous consumption of hot
fluid by geothermal power plants. A 100 MW power plant operated
by geothermal steam consumes between 1000 and 2000 t of steam/h.
This great amount of steam is taken from the subsurface and it
must be replaced by a natural influx of groundwater into the
hyperthermal reservoir. To avoid the cooling of the deposit, the
cooled fluids from the turbines are usually injected into rock
beds having no direct connection to the reservoir.

To classify hydrothermal deposits with high enthalpy, one needs
to distinguish between geothermal deposits containing predominately liquid, and geothermal deposits containing predominately
steam.

A "liquid dominated geothermal deposit" will be formed when water
or brine circulates in a reservoir with high permeability and
heat from an underlying heat source formed by a recent magmatic
intrusion is transferred into this reservoir. In such systems,

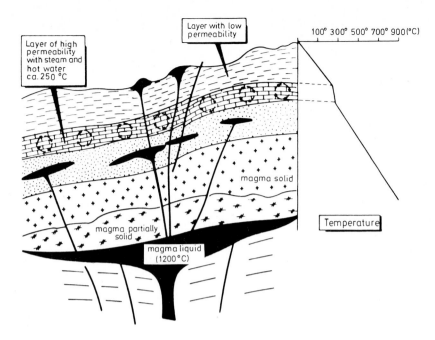

Fig. 6. Model of a natural steamfield near a volcanic intrusion — a high enthalpy deposit

temperatures of up to 300°C at depths down to 2000 m (Salton Sea hot-brine system) have been determined. At present, the most important, predominately liquid systems used for the generation of electricity are: Wairakei New Zealand with a capacity of 200 MW, Broadlands; Cerro Prieto (Mexico) 150 MW, Salton Sea (California); Yellowstone (Wyoming), Ahuachapan (El Salvador) 60 MW.

Economically, the "vapor dominated geothermal deposits" are the most valuable geothermal deposits. Superheated dry steam, often containing low percentages of carbon dioxide, hydrogen sulphide, and ammonia, can be produced from the hyperthermal reservoir. Geothermal reservoirs containing predominately steam are used for generating electric power in the following places: The Geyers (California, U.S.A.) 670 MW; Larderello (Italy) 400 MW; Matsukawa (Japan) 22 MW; Monte Amiata (Italy) 22 MW; Valles Caldera (New Mexico, U.S.A.).

The following types of geothermal deposits contain theoretically enormous, virtually inexhaustible amounts of heat. However, they cannot be exploited at present, because the necessary techniques required for this have not been developed yet.

"Hot magmatic systems", because of the large amount of energy they contain, are attractive for developing techniques for producing and using this energy. Basic data is already being collected in Hawaii with the objective of producing energy from hot magmatic systems that are connected with volcanoes. The active volcanoes in Europe (Vesuvius, Etna, Stromboli), volcanic regions such as the Canary Islands and the island of Melos are also suitable areas for research studies. The depths necessary for boreholes for developing such geothermal deposits would be in the range of 3000 and 6000 m with expected temperatures around 800°C (silicic magma) to 1200°C (basaltic magma).

Natural "hydrothermal systems" and "hot magmatic systems" contain only a small amount of the total heat in the earth's crust. In addition, they are restricted to relatively small zones with anomalous volcanic and tectonic activity. The majority of the underground is formed by rock beds with low permeability and little porosity. In depths of more than some thousand meters this rock is also hot. A vast amount of heat is stored in this rock, enough energy to cover the needs of mankind for many centuries. The most important aspect of this energy source is the fact that impermeable hot dry rock can be found in all parts of the world regardless of the geographical location. The heat from a rock body with a volume of 1 km^3 cooled by 10°C is 2.25×10^{16} J or 6.26×10^6 MWh which corresponds to an average output of about 25 MW over a lifespan of 30 years.

The exploitation of the thermal energy stored in this "hot dry rock system" requires special techniques for creating artificial heat transfer systems in several thousand meters depth. The general principle in such "man-made geothermal energy systems" is the transfer of heat from a hot dry rock reservoir to a fluid along artificially created heat-exchange surfaces. Because the thermal conductivity of the rock is low, very extensive surfaces are necessary for the heat exchange required for producing commercial quantities of heat for space heating and the generation of electricity.

For a number of years, theoretical and experimental investigations on this subject have been carried out at Los Alamos, U.S.A. and also in individual countries of the European Community. The results of these investigations show that man-made geothermal energy systems are technically feasible and that they are at or close to the point of economic feasibility at locations where the geothermal gradient is anomalously high.

Geopressured thermal deposits are formed by compaction of sedimentary materials for geological time periods in deep sedimentary basins containing alternating beds of sandstone and claystone. It is assumed that water is squeezed out of the soft marine sediments during sedimentation and then deposited in permeable sand/ sandstones. The content of the pores in these permeable layers is compressed to pressures much greater than the hydrostatic pressure. Thus, the geopressured water in the pores can be compressed to a pressure as high as the lithostatic pressure. The temperature of the pore water correlates with the geothermal gradient that results from the terrestrial heat flow and the thermal conductivity of the rocks. High methane concentrations at saturation, sometimes supersaturated, occur at these high pressures.

Geopressured thermal deposits have been identified in deep layers of young, rapidly sinking geosynclines, for instance in the Gulf of Mexico. These deposites have a very important energy potential, even in a place where the terrestrial heat flow is normal. Typical daily production from such a deposit would be several million m^3 of liquid containing many million m^3 of natural gas. In accordance with the great depths at which the development of these deposits is to take place, the temperature of the water ranges between 150°C and 180°C. However, we have not yet been able to overcome the tremendous technical difficulties in handling the high pressures and temperatures at depths of 5 to 10 km.

3 Environmental Implications

A problem not to be overlooked during the production of natural steam from zones of volcanic activity is the content of admixtures that could be of disadvantage for the human environment. Especially the release of CO_2, H_2S, and SO_2 could impair the quality of the human environment. Due to the usually low efficiency of geothermal power plants, there are also great amounts of waste heat, the disposal of which is problematical in densely populated regions. Additionally, the continuous extraction of water or steam from the subsurface has caused the subsidence of the earth's surface by several decimeters. The most troublesome problem is that surface waters and groundwaters could be contaminated by mineralized, gas-containing water from the plant.

Also, it seems to be important to investigate possible changes in the environment that could result from extraction of large quantities of terrestrial heat from the dry, low-permeable subsurface. Two phenomena should be studied in connection with this.

1. During the generation of artificial fractures, the natural field of tension is disturbed in the subsurface. Thus, earthquakes could be caused in seismically unstable zones. Such highly unstable conditions are extremely rare. Under these conditions, the energy involved in creating the artificial fractures (very small compared to the energy of an earthquake) will cause an earthquake somewhat earlier than it naturally occurs.

2. During long-term extraction of heat from the subsurface, contraction due to the cooling of the rock takes place. This causes a change in the state of tension. It is hoped that this contraction enlarges the heat-exchange surfaces. But it is possible that unwanted changes are caused by it in the subsurface and on the surface of the earth. However, the data obtained in connection with a small-scale, short-term heat extraction from impermeable rock complexes in Los Alamos, New Mexico, indicate no ecologically harmful effects.

Demands and Resources of Energy in the Present and Future

U.Lantzke and E.Meller[1]

1 Preface

Future energy supply prospects have changed drastically in the
70s and particularly over the past 2 years. The 1972-74 oil
crisis had for the first time made the oil consuming countries
aware of the danger of their considerable dependence on oil.
But once the oil flow was running smoothly again and oil prices
had stabilized, it was believed both that there was adequate
time to make the structural adjustments of energy economies and
that a satisfactory rate of economic growth could be achieved
in the 80s.

These prospects, however, were altered by the events of 1979 and
1980. The 1979 oil crisis was not simply a crisis of oil supply
and prices. It rather heralded an adjustment crisis whose effects
will be constantly felt in the years to come.

Future oil production prospects have changed fundamentally and
irretrievably. Some of the reasons lie in the events in the
Middle East and also — and partly conditioned by them — in OPEC's
more restrictive production policy, although at the moment,
i.e., spring 1981, the problem is partly hidden by the sharp fall
in oil consumption in the industrialized countries. The example
of Iran has shown that the development problems of the producer
countries cannot be solved in one generation. Many have therefore
slowed their industrialization plans and are instead contemplat-
ing making their oil reserves last longer. Since November 1978,
about 7 Mbd (million barrels per day; 1 Mbd = 4.86 Mtoe a year)
of oil production capacity has been withdrawn from the world oil
market. Crude oil production in the fourth quarter of 1980 was
about 6-7 Mbd below the 1979 figure. These figures clearly illustrate the reduced flexibility of the world oil market to which
consumer countries will have to accustom themselves in the future.
As a result of this, the time available for structural adjustment to reduced dependence on oil has been shortened. Even if the
net oil imports of the IEA countries fall further in 1981 and
possibly never again reach the 1979 level, because of the conservative production policies of the OPEC countries and growing oil
requirements in the rest of the world, above all in the developing countries, the world oil market will get tighter and tighter
during the 80s.

1 Internationale Energieagentur der OECD Paris, 2 rue André Pascal,
  75775 Paris Cedex 16, France

The energy future has become more uncertain and less predictable. Energy policy-makers can no longer reckon on a scenario free of surprises. The ability to anticipate unexpected events and prepare for them in advance is becoming increasingly necessary at both the national and international levels. The revolution in Iran and the subsequent conflict between Iran and Iraq have made it clear that the world will be walking a dangerous tightrope during the 80s. If it does not succeed in maintaining its equilibrium, serious economic and social upheavals will be unavoidable, to say nothing of the possible geopolitical consequences.

The following contribution on future energy demand and supply prospects is to be seen against this background.

2 Present and Future Energy Trends

What follows focuses more particularly on the energy situation in the industrial countries belonging to the IEA [1] which together account for about 80% of world primary energy consumption and over 70% of world oil consumption (the totals in each case exclude the planned economy countries).

Table 1 shows past and estimated future trends in energy supply and in particular those in world oil consumption and supply over the period 1973-1990. Estimated future trends are based on a synthesis of the forecasts made in mid-1980 by the IEA countries. Like all forecasts they are subject to constant change as a result of economic and political developments and the picture they give is true only of the relevant moment in time. In combination with past developments they nevertheless show important trends.

Since 1973 substantial progress has already been made in restructuring the energy economy in the direction of less dependence on oil, e.g.:

primary energy requirements have increased by only 1.2% a year whereas the economic growth rate has been 2.6% a year;

the rate of increase in oil consumption has slowed considerably, rising by only 0.6% a year;

overall energy efficiency (expressed as the ratio between primary energy requirements and GNP) has improved by 7%;

the production and consumption of alternative energy sources has increased by 11%, not least due to the tripling of the nuclear energy contribution.

As a result of these developments, 1979 oil imports were only slightly above the 1973 level, and if the quantity used in 1979 to replenish oil stocks to an above average level are deducted, 1979 oil imports were even below those of 1973. It is expected that this trend will continue over the next decade. It is therefore assumed, for example, that:

primary energy requirements will increase by about 2% a year against an economic growth rate of about 3.5% a year;

Table 1. Historical and projected trends, 1973-1990 (IEA, Country Projections)

	1973	1979	1985[a]	1990[a]
IEA (Mtoe)				
Total primary energy	3366	3612	4099	4591
Non-oil energy consumption	1618	1797	2204	2760
Oil consumption	1748	1815	1895	1831
of which:				
Net oil imports [b]	1173	1207[c]	1304	1265
World oil consumption (Mbd)				
IEA [b]	36.8	38.7	40.7	39.5
Other OECD	2.8	2.8	2.7	2.5
Others (including OPEC)	8.2	10.2	14.6	19.7
World (excluding CPE)	47.8	51.7	58.0	61.7
Non-OPEC supply (Mbd)				
IEA/OECD	13.9	14.9	14.1	13.6
Developing countries	1.9	5.2	8.5	11.0
CPE net exports (imports)	1.0	1.1	0.4	-1.1
Total	16.8	21.2	23.0	23.5
Additional requirements	31.0	30.5	35.0	38.2
OPEC production (Mbd)	31.5	31.6	?	?
Memorandum items				
IEA net oil imports (Mbd) [b]	24.1	24.8	26.8	26.0
TPE/GDP (Mtoe per 1975 $US)	0.89	0.83	0.77	0.72
Oil consumption as % TPE	51.9	49.8	46.2	39.9
Growth rate of GDP				
1979-85 (% p.a.)		–	3.5	–
1985-90 (% p.a.)		–	–	3.6
IEA oil production (Mtoe)	658	704	675	652
Non-oil energy consumption				
Coal	686	726	915	1156
Gas	684	713	762	845
Nuclear	43	125	253	398
Hydro	205	231	257	292
Others	–	2	17	69
Total	1618	1797	2204	2760

[a] Preliminary estimates made by IEA member countries. Non-IEA figures are Secretariat estimates

[b] Including international marine bunkers. Conversion factor: 1Mbd = 48.6 Mtoe/year

[c] Includes 53 Mtoe used to increase stocks

 oil consumption will at first fall slightly, over the next 5 years (by about 0.7% a year), then slowly increase again, so that the 1990 level will be only slightly above that of 1979;

 the oil share in total energy consumption will fall from 50% to 40%;

 energy efficiency will increase by a further 13%;

the consumption of alternative energy sources will increase
by over 50%, the greater part of the increase coming from coal
and nuclear energy.

Although these forecasts assume a substantial restructuring of
the energy economy of the IEA countries over the next 10 years,
there still remains an oil requirement of about 38 Mbd in 1990.
It is extremely doubtful whether the OPEC countries will be able,
and willing, to make such a quantity available to the world market.

Seen against this background, the magnitude of the challenge confronting the industrial countries is clear. If the energy policies of these countries do not considerably step up and accelerate efforts to reduce dependence on oil, the world oil market
and hence oil prices will come under increasing pressure. Under
these circumstances, sudden price jumps are more likely than
gradual rises. In addition, there is the danger that consumers
will be forced into a unilateral reduction of energy and oil consumption through a combination of higher prices and restricted
economic growth. Remedies to avoid or cushion the effects of such
a scenario do exist, however: greater efforts to make rational
use of energy, with particular reference to oil consumption, the
development of new energy technologies, increased use of coal
and nuclear energy, greater efforts to maintain oil production
and develop unconventional oil sources, increased trade in and
consumption of natural gas. The real difficulty in implementing
these remedies is actually that with few exceptions the problems
are not so much of a technical or of a financial nature, but lie
at the social, institutional, and political levels.

3 A Plausible Energy Future

The IEA is currently drawing up a set of plausible energy balances for the years 1990 and 2000. The provisional findings are
shown in Table 2. These energy balances are not forecasts, but
pointers intended to show the potential available for making
fundamental changes in the energy structures of the industrial
countries over the next 20 years, taking into account geological,
economic, and technological factors. Putting forward such pointers is intended to show the range and extent of the energy policy
measures which will be necessary to achieve a substantial reduction in oil consumption by the end of the century.

These scenarios are based on an estimated increase in energy
consumption of about 1.6% a year over the next 10 years and about
1.8% thereafter. Under these conditions it would be possible,
making great efforts to use energy rationally, to achieve an annual economic growth rate of 3% to 3.5%. As a result, the energy
consumption of the IEA countries in the year 2000 would amount
to about 5100 Mtoe as compared with 3600 in 1979. The question
is: by what means, in realistic terms, can this demand be satis-
fied?

  Coal's share in primary energy requirement coverage could be
  increase from the present 20% to 27% by 1990, and 35% by the

Table 2. Plausible energy scenarios, 1990 and 2000 (IEA)

	1979	("Plausible" scenarios) 1990	2000
IEA (Mtoe)			
Total primary energy	3612	4300	5140
Non-oil energy consumption	1797	2670	3810
Oil consumption	1815	1630	1330
of which:			
Net oil imports [a]	1207	1010	730
IEA oil production	704	700	700
Non-oil energy consumption			
Coal	726	1160	1815
Gas	713	850	1000
Nuclear	125	340	625
Hydro/Geothermal	231	290	320
Others	2	30	50
Total	1797	2670	3810
Memorandum items			
IEA net oil imports (Mbd) [a]	24.8	20.8	15.0
Oil consumption as % of TPE	49.8	37.9	25.9
Avg. annual growth TPE			
1979-1990 (% p.a.)	–	1.6	–
1990-2000 (% p.a.)	–	–	1.8

[a] Includes international marine bunkers

year 2000. As compared with 1978, this would mean that coal consumption would increase by 70% by 1990 and triple by the end of the century.

Nuclear energy's share in primary energy requirement coverage could be increased from the present figure of barely 4% to 8% by 1990, and 12% by the year 2000.

Oil production in IEA countries could be maintained at the present level.

Natural gas consumption could be increased overproportionally, primarily by means of increased imports, which could reach about 6 Mbd by the year 2000.

Hydraulic power and geothermal energy could be increased by almost 40% over the next 20 years.

Alternative energy sources, in particular solar energy and energy from the biomass, could contribute about half as much as nuclear energy does today by the end of the century.

The above scenario is necessarily only illustrative and at this stage merely indicates directions in which energy policy has to work. If the path shown, which is in fact practicable but needs increased efforts, were to be followed, the oil share in total energy consumption would fall from about 50% to 25% by the year 2000. Net oil imports would fall from the current 24 Mbd to 15 Mbd by the year 2000. Oil could be increasingly reserved for the transport sector and non-energy sectors and electricity's share in meeting energy requirements would grow from the present 32% to over 40% toward the end of the century.

Pursuing such a scenario would have the advantage of making our energy supplies more secure over the next two decades, creating the basis for an adequate rate of economic growth and giving the developing countries a chance of importing their oil needs, which are tending to increase, without putting too much strain on the stability of the world oil market.

4 Making the Transition

It will not be easy to achieve such a change and adjustment of the energy economies of the industrial countries. Awkward and often unpopular decisions will be necessary in many countries.

4.1 Rational Energy Use

More rational energy use and increased oil substitution will be the cornerstones of such a policy. In view of the long lead times involved in increasing the supply of energy, whether through nuclear power or new energy technologies, conservation and rational energy use will have to be given the highest priority in the short and medium term. Only with their help will it be possible to bridge the gap which can be predicted for the second half of the 80s between energy supply and demand without serious economic friction. Energy prices will be at the center of such a policy. But this does not exclude additional supportive measures by governments to accelerate market processes and stimulate economic transactors to take special measures. This also involves actions at the international level. In October 1980, the IEA started the Energy Management Initiative [2] in all member countries and on the occasion of the Ministerial meeting in December 1980, 19 specific fields of action were adopted as priorities in drafting national energy policies.

Figure 1 shows the ratios between primary energy requirements and economic growth and between primary energy consumption and oil consumption in recent years, and expected developments to 1990.

4.2 Oil

Table 3 shows remaining oil reserves, cumulative production, and estimated additional recoverable resources. Total world oil consumption in 1979 amounted to 3100 million tonnes, i.e., less than 1/30th of the remaining proved reserves. The great dependence on the Middle East (almost 60%) is obvious. Altogether, the OPEC countries account for almost 70% of proved reserves.

As can be seen from the high ratio of cumulative production to proved recoverable reserves, production of (conventional) oil in North America has probably passed its peak and production in this area will further decline over the next few years. In order

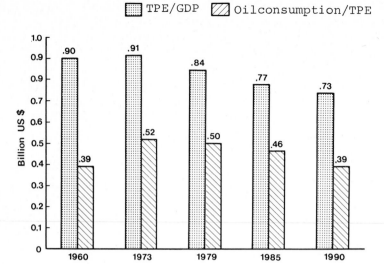

Fig. 1. Energy demand ratios (IEA countries)

Table 3. World ultimately recoverable resources of conventional oil as at 1 January 1979 (million tonnes). (Survey of Energy Resources, 1980 [3])

	Cumulative production	Proved recoverable resources	% of total	Est. additional recoverable resources	Total
Africa	3,750	8,040	9.0	34,000	45,790
North America	17,520	4,480	5.0	24,000	46,000
Latin America	7,040	7,770	8.7	12,000	26,810
Far East/Pacific	1,720	2,390	2.7	12,000	16,110
Middle East	14,680	51,050	57.3	52,000	117,730
West Europe	560	2,710	3.0	10,000	13,270
U.S.S.R., China, East Europe	7,530	12,700	14.2	64,000	84,230
Antarctic	–	–	–	4,000	4,000
World	52,800	89,140	99.9	212,000	353,940

to maintain oil production at the current level in the IEA area it is therefore necessary to make great efforts in the exploration and development of new oil fields, above all in the North Sea and the North American frontier areas where promising discoveries have recently been made. In addition, increasing reliance will have to be placed on enhanced oil recovery and unconventional oil, chiefly shale oil in the United States and Australia, and oil sands in Canada.

4.3 Coal

Table 4 shows world coal reserves. Current world coal consumption is slightly over 2500 Mtoe of energy, or about 1/260th of the technically and economically recoverable reserves.

Table 4. Coal resources and reserves (million tonnes coal equivalent). (World Energy Conference and Report of the World Coal Study)

	Technically and economically recoverable reserves	% of total	Geological resources	% of total
Australia	32,800	4.9	600,000	5.6
Canada	4,242	0.6	323,036	3.0
Federal Republic of Germany	34,419	5.2	246,800	2.3
United Kingdom	45,000	6.8	190,000	1.8
United States	166,950	25.2	2,570,398	23.9
India	12,427	1.9	81,019	0.8
Republic of South Africa	43,000	6.5	72,000	0.7
People's Republic of China	98,883	14.9	1,438,045	13.4
Poland	59,600	9.0	139,750	1.2
Soviet Union	109,900	16.6	4,860,000	45.2
Other Countries	55,711	8.4	229,164	2.1
Total world	662,932	100.0	10,750,212	100.0

The world coal study [4] published last year assumes that coal consumption in the OECD area could rise to 2000-2025 Mtce by the end of the century. The upper figure corresponds to 1950 Mtoe, while the IEA plausible energy scenario (see Table 2) assumes 1815 Mtoe. A total coal consumption of this order of magnitude would require a considerable increase in coal trade which would reach almost 1000 Mtce by the year 2000. Possible world coal trade movements in the year 2000 are shown in Table 5.

Increased coal utilization has become an attractive option for practically all industrial countries, but especially for the United States. At the Venice World Economic Summit in 1980, the seven biggest industrial countries decided to double coal production and use by 1990. Greater efforts than hitherto will certainly be required if this ambitious target is to be reached. Among other things more speed and determination than has been forthcoming up to now will be necessary in converting existing oil-fired power stations to coal and building new coal-fired stations.

In May 1979, in order to encourage coal trade and use, the governments of the IEA member countries decided on an outline program for coal and in April 1980 they established a Coal Industry Advisory Board made up of 33 high-level representatives of the coal industry, i.e., producers, traders, and users. A first report containing detailed suggestions for increasing coal production and use was presented to the IEA energy ministers in December 1980 [5].

Table 5. Hypothetical trading preferences for sources of coal supply — Year 2000 (Mtoe). (Report of the World Coal Study)

Importer Country/Region	Exporter source country															
	Low case								High case							
	A	B	C	D	E	F	G	Total sources	A	B	C	D	E	F	G	Total sources
Denmark	1	–	2	3	1	–	2	9	2	2	6	5	2	1	3	21
Finland	–	1	–	5	1	–	2	9	1	2	–	5	2	1	2	13
France	7	4	8	5	3	1	10	38	28	24	20	7	19	1	16	115
Germany, Fed.Rep.	6	5	3	2	1	2	1	20	12	10	5	4	2	4	3	40
Italy	7	6	4	4	–	1	6	28	16	9	12	8	–	3	10	58
Netherlands	6	2	6	6	1	–	2	23	10	5	10	11	1	–	1	38
Sweden	4	2	–	4	3	–	4	17	7	3	–	6	5	1	4	26
United Kingdom	1	–	–	–	–	–	1	2	12	–	–	5	–	–	–	17
Other Western Europe	11	8	9	12	3	–	12	56	18	13	12	11	6	2	12	74
OECD Europe	43	28	32	41	13	4	40	200[a]	106	68	65	62	37	13	51	400[a]
Canada	–	17	–	–	–	–	–	17	–	9	–	–	–	–	–	9
Japan	45	43	5	1	18	10	10	132	78	66	6	1	25	16	14	206
East and other Asia	34	23	15	–	20	4	4	100	86	52	33	–	48	–	8	227
Africa	1	1	–	–	–	–	1	3	12	10	–	–	5	–	3	30
Latin America	18	13	–	6	6	–	17	60	18	13	–	6	6	–	17	60
Centrally planned economies	–	–	–	25	–	–	25	50	–	–	–	25	–	–	25	50
Total world [a]	140	125	55	70	55	20	95	560	300	215	105	95	120	30	115	980

[a] Totals are rounded

Legend: A = Australia; B = United States; C = South Africa; D = Poland; E = Canada; F = People's Republic of China; G = Other

4.4 Natural Gas

The IEA plausible energy scenario (see Table 2) assumes that
natural gas consumption will increase more or less in parallel
with total energy consumption and thus maintain its present
share of nearly 20%.

As an energy source, natural gas has a number of advantages. It
has a reputation as an ecologically clean energy source and does
not have to overcome psychological barriers as is the case with
coal or nuclear energy, for example. The technologies for its
production and use are known and the necessary infrastructures
are to a large extent available both in Europe and in North America. The critical points are the availability and price of natural gas. Table 6 shows that natural reserves are considerable
and are distributed over a relatively large number of areas. In
1979 total natural gas consumption amounted to about 1.8 trillion
cubic meters (tr m^3) or about 1/50th of proven exploitable reserves.

The reserves alone would allow present world trade in natural
gas to be multiplied several times over and the locations of
about 23 tr m^3 (over 30% of total) are conducive to such an expansion of international trade in gas [6]. It must be noted,
however, that of this quantity about 12.5 tr m^3 are in the Soviet
Union, 5.3 tr m^3 in Iran, and about 3.7 tr m^3 in other OPEC countries. The relatively high costs of gas storage and transport
will at first hinder the development of the more remotely situated reserves. Whether the enormous gas reserves will be able to
be developed and used will depend to a large extent on the economic and political conditions and relations within and between
producer and consumer countries. Critical factors here are price
policy, considerations regarding the security of national energy
supplies, and the question of the relative economics of pipelines
versus LNG transport. Events in 1980, such as the negotiations
between Algeria and American and Western European gas companies
and the highly controversial project of a gas agreement between

Table 6. World ultimately recoverable resources of natural gas, as at 1 January, 1979 (trillion cubic meters). (Survey of Energy Reserves 1980 [3])

	Cumulative production	Proved recoverable resources	% of total	Est. additional recoverable resources	Total
Africa	0.1	7.3	9.9	26	33.4
North America	16.9	7.5	10.1	42	66.4
Latin America	1.8	4.7	6.3	10	16.5
Far East/Pacific	0.2	3.3	4.5	10	13.5
Middle East	1.1	20.5	27.7	30	51.6
West Europe	1.5	3.9	5.3	6	11.4
U.S.S.R., China, East Europe	5.2	26.9	36.3	64	96.1
Antarctic	–	–	–	4	4.0
World	26.8	74.1	100.1	192	292.9

Western European companies and the Soviet Union have made this clear. It is, however, to be expected that trade in gas, above all by sea, will expand considerably, mainly in the Far East, where Japan intends to triple its imports by 1990.

4.5 Nuclear Energy

The most uncertain factor in the IEA plausible energy scenario (see Table 2) is beyond doubt the forecast concerning the nuclear energy contribution. This is put at about 7 Mbdoe by 1990 and 12-13 Mbdoe by the year 2000, equivalent to almost half of current OPEC oil production. In order to achieve this goal all nuclear power stations now under construction or on order need to be completed and new power stations of about 20,000 MW will have to be ordered each year over the next 10 to 12 years. Such a program, though technically and economically feasible, will require a spectacular reversal of the current stagnation of nuclear energy development. After a decade during which the nuclear energy contribution tripled and achieved an average annual growth rate of over 25%, the world nuclear energy contribution increased by only 3% in 1980. In the United States no new nuclear power stations were ordered, on the contrary many projects were canceled. In Germany, no first construction licences have been issued since the mid-70s and only one new order has been placed. The only extension program that has so far gone ahead without disruption is France's nuclear energy program. There, over the next few years, one nuclear power station will come on stream every 2 months. In some other countries too (e.g., Japan), signs of a revival and acceleration of the nuclear energy program can be seen in the wake of the latest oil crisis. In order to achieve the extension of nuclear energy desirable on energy policy grounds, it will be important to enable public opinion, by means of objective information and discussion, to make the choice between the minimal risks involved in the use of nuclear energy and the risk, in our opinion far from hypothetical, of not having enough energy. In addition, it will be necessary to fully exploit other energy options, in particular energy conservation, and resolve the remaining problems of using nuclear energy, in particular waste disposal, by experimental demonstration and use of the solutions technically available.

Table 7 shows estimated world nuclear installed capacity for 1985 and 1990 excluding the planned economy countries. By the year 2000, a capacity of 476 to 549 GW can be expected in the OECD countries and from 626 to 735 GW in the world (excluding the planned economy countries). The latter figure is considerably below the INFCE Conference [7] estimate of 850-1200 GW by the year 2000. However, INFCE came to the conclusion that proven natural uranium reserves would be sufficient to meet fuel requirements into the next century. This would give sufficient time to develop fast breeder reactors, which use the available uranium 60 times more efficiently than the present light-water reactors, to the commercial stage and thus broaden the nuclear energy option.

Table 7. Forecast of Free World nuclear power capacity (GW at year end)

	1980	1985	1990
North America	59.3	101.8	136.9-141.8
United States	53.8	90.0	121.0-125.0
Canada	5.5	11.8	15.2- 16.8
Pacific	15.0	21.0- 28.0	28.0- 47.0
Japan	15.0	21.0- 28.0	28.0- 47.0
Europe	43.8	85.3- 93.4	102.6-139.9
France	14.5	29.9- 35.4	36.7- 55.2
Germany	8.6	17.0- 18.3	19.3- 24.5
Italy	1.1	1.1- 2.4	3.4- 5.4
Spain	1.1	7.5	9.5- 14.5
Sweden	5.5	8.3	9.4
United Kingdom	7.5	10.6	13.4- 14.8
Other	5.5	10.9	10.9- 16.1
OECD total	118.1	208.1-223.2	266.8-328.6
Non-OECD, non-CP		14.0- 16.0	24.0- 33.0
Free World total		222.1-239.2	290.8-361.7

Note: The projections for IEA countries are generally lower than the official
      country projections. Rough estimates for installed nuclear capacity in
      2000 are between 476 GW and 549 GW for the OECD, and between 626 GW and
      735 GW for the total Free World. Figures for non-OECD member countries
      are those of the INFCE estimates

4.6 New Energy Sources and Technologies

It is difficult to estimate the contribution of new energy
sources to future energy supply. Although a great deal of technical research and policy attention has been devoted to new energy sources in the last decade, particularly after the first
oil crisis in 1973/74, the contribution of renewable energy
sources such as solar, wind, geothermal, biomass (including al-
cohols), and mini-hydro, to total energy balances is still insignificant. However, some of the contribution is not overtly
indicated; for example, in the case of passive solar and heat
pumps, it is recorded as a reduction in energy demand, rather
than a new source of supply.

The share of new energy sources could reach 5% to 10% of industrialized countries TPE by the end of the century, varying, of
course, considerably from one region to another. Whether this
contribution will be achieved depends not only on factors such
as financial resources devoted to the research and development
of new energies and technologies but also on the price of oil,
the efforts to commercialize such energies and the removal of
institutional impediments in some countries.

5 Concluding Remarks

Possibilities for assuring balanced energy supplies in the future do exist. The resources and technologies for gradually replacing oil by alternative energy sources are available. The risk of a further oil crisis, however, with possibly even more serious consequences for the economic and social structures of the industrial countries and the rest of the world can certainly not be excluded. But through the determined implementation of an energy policy reducing oil consumption to the utmost and making real use of all other energy sources, the risk can be reduced and its consequences kept within calculable limits.

References

1 The International Energy Agency, founded by the industrialized countries in 1974 in response to the first oil crisis, includes all OECD member countries with the exception of France, Finland, and Iceland. The IEA secretariat is currently preparing a world energy outlook which is expected to be published in spring 1982
2 IEA, energy management guide. OECD Paris, September 1980
3 Rahmer BA (1980) The world energy potential. Petrol Econ 11/80:479-483
4 Report of the world coal study (1980) Coal — bridge to the future. Ballinger Publ Comp, Cambridge Mass
5 IEA, report of the IEA coal industry advisory board. OECD, Paris, December 1980
6 Jensen Associates Inc (1979) "Imported liquified natural gas". A report to the Congress of the United States. Office of technology assessment. September 1979
7 Internat nuclear fuel cycle evaluation conference (1980) Summ vol, IAEA, Vienna

Energy Strategies

P.J.Jansen[1]

1 The Energy Problem

1.1 In the Industrialized Countries

The energy problem in Industrialized Countries is threefold:

First, it is a price problem.

The rise of crude oil prices caused a trade imbalance in most of the industrialized countries. Severe economic and political consequences are expected and a worldwide economic crisis may develop.

Second, it is a problem of energy availability.

It has become apparent that the world or at least a major part of the industrialized countries could suddenly be confronted with an interruption of a significant amount of the oil supply. The danger of wars initiated as a result of this is increasing.

Third, it is an acceptance problem.

The difficulty of getting people to understand the measures necessary for solving the energy problem grows continuously; political consensus on the subject is rarely achievable. The controversy has already taken on the character of confrontation. Political stability could earnestly suffer the consequences.

The roots of these three problems are deep and interwoven:

The price problem and the problem of energy availability only signal the tip of a problem field, concerning the relations between the industrialized countries and the Third World. The first, and possibly one of many future weapons, to be used by the Third World, is OPEC and its oil, which should force the industrialized countries to take notice of the dichotomy of receiving substantial supplies from the resources of the Third World on the one hand, and supporting the development in the Third World through improved living conditions on the other.

The acceptance problem stems very much from the worldwide uneasiness about the end to which industrialization and eco-

---

1 Technische Universität Wien, Institut für Energiewirtschaft, Wiedner Hauptstr. 7, 1040 Wien, Austria

nomic growth will lead. Too often single interests dominated public ones; too often policies were inconsistent. How can the public be convinced in the face of these to accept the incongruity that certain actions are necessary to resolve future problems, if those actions personally affect their style of living.

1.2 In the Third World

The energy problem exists quite similarly in the Third World:

64 developing countries, including one of the poorest, imported 75% of its commercial energy demand, that is, oil [1]. The extreme price increase in oil brought all these developing countries nearly to insolvency and deprived them of the chance for economic development.

According to a World Bank analysis [1], it is estimated that still 50% to 65% of the energy demand in Asia and 70% to 90% in Africa is met by using lumber and agricultural waste. The deforestation rate is 1.3% per year. Already 46 nations are faced, or nearly faced, with supply problems, a local energy problem of worldwide importance.

The economic development of the oil-exporting countries cannot cope with the current level of oil export. Severe social tensions would have to be expected if one could enforce an equivalent import level. In case of slower development either the oil is underpaid — as in the past, or by worldwide inflation — or the oil export must be reduced.

The roots of these problems, too, originate in the industrialized countries' unlimited demand for raw materials and their inability to resolve the Third World's problems cooperatively and successfully.

1.3 As a Growth Problem

... *of Economics:* The energy problem is even more apparent if one looks at the following facts.

Table 1. Per capita income [2]

	GDP per capita 1975 (in 1975 $)	World population 1975 (billion people)
Industrialized countries	4600	1.2
Developing countries	1200	0.5
Least developed countries	300	2.3
World average	1700 at	4.0

Table 1 shows that more than half of the world's population lives on less than one-tenth of the per capita income in industrialized nations.

Since 1950 the GDP growth rates had been 3.6% per year and per capita on a world average. With a 6.7% rise the U.S.S.R. and Eastern Europe have made the step, however, from $ 700 per capita in 1950 to an industrialized country's level of $ 3400 per capita in 1975. A similar step was performed by the Middle East and North Africa from $ 350 to $ 1400 per capita. At the same time, the United States, which was in 1950 already at a level of $ 4400 (in 1975 value) per capita, grew only by 1.9% per year and per capita to a level of $ 7000 per capita in 1975, a level it shares with the Federal Republic of Germany.

Nevertheless, even the industrialized nations will go on growing economically, or at least want to, in order to guarantee social and economic stability. It is generally considered that countries like the United States or the Federal Republic of Germany reach a plateau with a lower limit of about $ 14,000 per capita within the next 50 years [2,3]. This would correspond to an average economic growth rate of 1.4% per year and per capita only.

... *of Population:* The future not only requires a giant advance in the living standards of the majority of the world's population to catch up with the industrialized countries, but brings forth 8 billion people within the next 50 years out of 4 billion, main- ly due to a birth surplus in the least developed countries [2]. If we are lucky, this may be an upper limit. Whether it is 8 or 10 billion in the end, is not the point. However, in order that higher multiples of population explosion should not occur, social and economic development plans must be implemented and must be successful.

Assuming that at some future time 8 billion people could realize $ 14,000 per year and capita (twice the present day United States amount), the present world economic activity level would be re- quired to rise 17.5 times. To reach that goal, the world's eco- nomy would have to grow by an average of 3% per year for 100 years, and 5% per year for the poorest countries which triple their population. These figures do not look completely unreason- able and one could well find them desirable.

... *of Energy:* It is not as easy, however, to imagine that the energy demand can and should grow 17.5 times.

The 1975 energy demand was 8.2 TWa (roughly 8200 million tonnes of coal equivalent per year).

Table 2 shows a breakdown into major energy resources. Oil and gas are summed up in order to show more clearly the energy de- pendence problem, since both resources may soon have similar availability problems.

With a complete change of the present way of thinking one could find ways of supplying 17.5 times this amount of energy. By sup- plying more or less all energy services by electricity and hydro- gen, produced by solar and nuclear power stations, 17.5 times as much as the present energy production level would correspond to a power park of 5000 MW electric equivalent per 600,000 people. In other words, this corresponds to 140 TWa per year (thermal)

Table 2. Energy supply 1975 (TWa per year)

Crude oil and gas	5.1
Hard coal and lignite	2.3
Nuclear energy	0.1
Renewables	0.7
World energy demand	8.2

for the whole world. But this is only playing with numbers and neglects the political realities, which have to be elaborated on further.

2   Energy Scenarios

2.1   The IIASA Analysis

It is necessary at this point to refer in more detail to the IIASA analysis "Energy in a Finite World" [2]. The analysis is one of the most detailed studies on world energy problems and its potential solutions, covering a horizon of 50 years. Much more realistically than could ever be done in this paper, the analysis shows the potential development paths of the world, broken down into seven regions, with an analysis of the energy demand of each of them and the possibilities and linkages of its supply.

2.1.1   *The High Scenario*

The so-called High Scenario aims at a rapid development of the Third World and, to achieve this, it is deemed necessary that the industrialized countries also have a prosperous economic development. The forecasts for the High Scenario are summarized in Table 3.

Table 3. High Scenario economic activity level [2]

	GDP in $ per capita in 2030	Factor with reference to 1975	World population in 2030 (billion people)
Industrialized countries	20,000	4.3	1.6
Developing countries	6,000	5.0	1.2
Least developed countries	1,000	3.3	5.2
World average	5,400	3.2 at	8.0

While today's energy needs are worldwide equal to 1.25 W per $, the High Scenario assumes energy conservation effects to the extent of reducing this ratio to 0.8 W per $ in 2030. This results in a worldwide energy demand to 36 TWa per year and is called the 36 TW Scenario. A detailed resource allocation analysis resulted in a most reasonable estimate for the energy supply pattern, given in Table 4.

Table 4. High Scenario energy supply pattern [2]

	TWa per year	Factor with reference to 1975
Crude oil and gas	12.8	2.3
Hard coal and lignite	12.0	5.2
Nuclear energy	8.1	
Renewables	2.8	4.0
World	35.7	4.3

The IIASA analysis explains what efforts are necessary to enhance oil and gas recovery to such an extent. The same is true
of coal. On the other hand, it is ultimately possible. The nuclear energy contribution corresponds to about half of what a
detailed study about the capital requirements and various steps
necessary on the fuel cycle side has found to be the absolute
upper limit for 2030. Still 8.1 TWa per year, or roughly 4000
GW-electric, require a big effort to be made within 50 years.

It is worth mentioning that more realistic judgements on expected structural changes in various societies and patterns of
growth conditions are far removed from the previously stated
vision of a 140 TW scenario. In relation to the aforementioned
North/South problem it is also important to realize that despite
a more than tripling of the per capita income in the developing
countries, the High Scenario widens the income gap between industrialized and least developed countries. The income difference primarily originates in the population explosion of the
Third World.

2.1.2 The Low Scenario

By introducing this paper with the worldwide energy problems,
it has become apparent how reasonable it is to believe that mankind's supply of energy may remain below 36 TW. Consequently,
the IIASA Low Scenario assumes lower economic growth rates and
further improved energy conservation. Table 5 shows the economic
activity levels of the Low Scenario.

Table 5. Low Scenario economic activity levels [2]

	GDP in $ per capita in 2030	Factor with reference to 1975
Industrialized countries	11,000	2.4
Developing countries	3,100	2.6
Least developed countries	600	2.0
World average	3,000	1.8

With slightly less than 0.7 W per $ one arrives at a 22 TW Scen-
ario broken down according to the IIASA proposal in Table 6.

It is worthwhile to go into more details of this scenario as it
still enables the least developed countries to double their per

364

Table 6. Low Scenario energy supply pattern [2]

	TWa per year	Factor with reference to 1975
Crude oil and gas	8.5	1.7
Hard coal and lignite	6.5	2.8
Nuclear energy	5.2	
Renewables	2.3	3.3
World	22.5	2.7

cápita income, with more than 5 billion people, but not sooner
than in 50 years. Of course, the energy demand level would still
be nearly three times that of today.

The economic growth for the United States and the Federal Republic of Germany up to the year 2000 only amounts to 1.5% to 2.0%
per year (against an average of all industrialized nations of
3% per year), halving beyond 2000. These figures are not improbable according to present knowledge, but one expects them accom-
panied by severe economic problems. In addition, one hesitates
to believe that the industrialized countries thus are able to
contribute substantially to the Third World's development.

The IIASA Low Scenario takes economic growth rates of about 4%
per year for the non-industrialized countries. Per capita the
growth rates are only 1.7% per year, which is only half the average since 1950.

In 1975 the United States, U.S.S.R., and China used about 1.5 W
per $, while Europe remained at 0.9 W and the Federal Republic
of Germany at 0.8 W per $. The 22 TW Scenario assumes that all
industrialized nations will reach 0.8 W per $ (Europe 0.7 W per
$). The developing countries were placed at 1.0 W per $ for 2030.
For the least developed countries the phase of starting industrialization (with priority to economic development as compared
with the rational use of energy) is characterized by an average
of 1.4 W per $ in 2030, comparable to historical data in indus-
trialized nations. On world average the energy demand thus drops
from 1.25 W per $ to an equivalent of 0.9 W per $.

3. Energy Conservation

3.1 The Enquete Commission of the German Parliament

In the Federal Republic of Germany a commission of parliamentarians and scientists with controversial views on the energy
future has analyzed the real options of an energy policy. Much
emphasis was given to the possible structural changes in the
economy, that is, saturation effects in certain areas of consumption, the growth patterns of industry related to energy intensive-
ness, and the chances of energy conservation. All in all, the
demand side of the energy problem was investigated in depth [3].
With 62 recommendations toward the more rational use of energy,
the analysis is certainly one of the most sophisticated for

Table 7. Energy paths of the German enquete commission

	Path 2		Path 3
$ per capita 1975		7,200	
$ per capita 2030 (in 1975 $)	21,000		21,000
W per capita 1975		5,800	
W per capita 2030	11,000		7,200
W per $ in 1975		0.81	
W per $ in 2030 (in 1975 $)	0.52		0.34

highly industrialized societies. It is impossible here to go
into details. Nonetheless, indices can be drawn from this analysis which are of interest within our global discussion.

The commission described four possible paths of energy policy
futures, two of which, path 2 and path 3, were adopted by the
commission's majority as the desirable options. It is believed
that they describe the range of realistic energy futures. Some
indicators of these two paths are shown in Table 7.

It should be mentioned that the high economic activity level in
2030 is due to an economic growth of only 2% per year. But the
main point is that a sophisticated use of technique is said to
correspond to 0.4 or 0.5 W per $.

3.2 The Zero Growth Scenario of IIASA

With this background one may more easily accept the 16 TW Scenario of IIASA [2], which is described as a variant in case of
extreme world hardship. The economic growth rates are unchanged
in the various world regions compared to the 22 TW Scenario.
Again the specific energy need in the non-industrialized countries remains unchanged. The specific energy need for the industrialized countries is reduced, however, to 0.5 per $, thus main-
taining the world's 2000 W per capita of today, even in the year
2030. The energy demand doubles as the population doubles. The
energy needed to improve the standard of living in the least
developed countries is a transfer from the industrialized countries caused by a more efficient use of energy including a shift
of economic growth toward less energy-intensive production sectors.

3.3 Two Ways to Meet the Demand

To meet 16 TWa per year the supply strategy may be quite different. It is interesting here to note that with constant world production of crude oil, gas, and coal the gap caused by the population explosion can be closed in principle either by nuclear
energy or by renewables.

If the renewable contribution is held to current levels, then
8 TWa per year of nuclear energy are needed, which roughly cor-

Table 8. Zero growth scenario energy supply pattern

	TWa per year	Factor with reference to 1975
Crude oil and gas	} 7.4	} 1.0
Hard coal and lignite		
Nuclear energy	4.5	
Renewables	4.5	6.4
	16.4	2.0

responds to 4000 GW electric equivalent worldwide or 5000 MW
electric equivalent per each 10 millions of people.

In a non-nuclear case, we need 8.6 TWa of renewables, which is
way below the maximal potential given in the IIASA analysis. It
is important, however, that about half of it is biomass and in
competition with world food production. It is reasonable, therefore, to diversify, which means using nuclear energy as well as
renewables. So assuming the economic activity levels of Table 5,
Table 8 shows how 16 TWa per year could be supplied.

3.4 Consequences

It must be pointed out that the 16 TW Scenario means an effort
in seven areas simultaneously:

1. keep the population growth under control down to a level of
   about twice as much people as today

2. address and cope with a reduced economic growth in the industrialized countries

3. manage continuous development in the Third World

4. achieve a higher energy efficiency in the industrialized
   countries

5. intensify the use of renewables on a global basis

6. get nuclear energy introduced and accepted

7. hold steady the availability of oil, gas, and coal supply

In comparing the previously discussed scenarios and realizing
what is meant by these efforts quantitatively in the 16 TW Scenario, one gets the feeling that failing in one of the seven efforts even partially will increase the goals for others in a
way which makes the crisis more probable. The author of this
article feels that the quantitative goals of the 16 TW Scenario
are best balanced for a minimal risk future. Unfortunately worldwide cooperation to get this strategy going is very unlikely.

Let us ask in summary, as fictitiously as at the beginning, which
per capita income at 0.5 W per $ for the whole world (8 billion
people) corresponds to 16 TW? It is $ 4000 per capita. If the
Third World should reach this level as well, this means zero
growth for the industrialized countries, and most probably a
step back for the richer ones.

3.5   New Orientation

The presentation of these quite different scenarios and their
assumptions with respect to

    population growth
    economic growth in industrialized
                        developing
                        and least developed countries
    energy conservation efforts
    and changes within the growth pattern of industry
    as well as successes in different energy supply
    technologies

have shown that a projection of the world energy demand and its
supply is impossible. The scenarios lack also the normative character so often present. The normative character means that desirable scenarios should be realized by political actions. Too
many of these parameters, however, are not really open to influence and can never be agreed upon on a global basis. Neither do
the instruments for adequate control or the political structures
for such worldwide consensus mechanisms exist. Last but not
least, the least developed countries add a completely unpredict-
able element by constituting a hungry multitude within a histo-
rically new situation.

4   Courage for Action

There is no way to speculate, therefore, whether mankind will
cope with the problems without disaster. The scenarios can show
us, however, that every effort which at least brings some improvement to one of the problems must be made. Supersensitivity
and debates about optimal solutions, such as debates whether nuclear energy or energy conservation is more appropriate, should
no longer be tolerated. We have to undertake all that is in our
power. The seven efforts listed in the 16 TW Scenario may be re-
considered here.

In the FRG commission [3] parliamentarians and scientists of
quite different opinions were able to develop a common energy
policy, at least for one decade:

    high energy conservation efforts
    and nuclear energy.

In reaction to what was said at the beginning with respect to
the acceptance problem, we hope that an open discussion of what
can be done politically and about the global energy problem may
eliminate controversies. One thing is clear despite the uncertainty in the scenarios: energy demand will grow significantly.

4.1   Cooperation with the Third World

The additional energy problems, mentioned at the beginning, must
be attacked by establishing a new relationship between the industrialized countries and the Third World. But this is no guarantee for success. It would lead us too far astray to mention

368

the complicated worldwide economic aspects. Nevertheless, it should be mentioned once again that for a long time OPEC has repeatedly called upon the industrialized nations to improve the present situation in a way that meets the Third World's desire toward a sound development [4].

What the Third World needs are robust and small technologies which fit into the social structures and support an intrinsic development. A detailed careful analysis is needed. Success would result in a twofold benefit. First, additional export stabilizes the aforementioned imbalance of payments. Second, the subsequently demonstrated cooperation eases the price pressure and reduces the danger of an interruption of oil supply or even of other raw materials.

The key to solving the energy problem is not so much related to energy technology but to economic and global policy. Energy strategies of the future are strategies of cooperation with the Third World, economic and political strategies to develop the Third World. In the context of worldwide strained relations this means a new unique effort in the diplomatic field, by behaving as a partner, as a friend making amends. It means a cooperative undertaking, a noble one rather than ruinous bargaining. If such a future is viewed as utopian, the alternatives are only dismal.

References

1 Energy in the developing countries, World Bank, Washington DC, August 1980
2 Häfele W et al. (1981) Energy in a finite world, vol I and II. Ballinger, Cambridge Mass., cited according to
   Gerwin R (1980) Die Welt-Energieperspektive. DVA, Stuttgart
3 Zukünftige Kernenergie-Politik (1980) Zur Sache 1 und 2, Dtsch Bundestag, Bonn
4 Servan-Schreiber JJ (1980) Die totale Herausforderung. Molden, Wien

Subject Index

J. S. Rinehart

Geysers and Geothermal Energy

1980. 97 figures, 43 tables. XVI, 223 pages
ISBN 3-540-90489-1

Contents: Geysers of the World. – The Geologic, Thermal, and Hydrologic State of the Earth. – Fundamentals of Geyser Operation. – The Role of Gases in Geysers. – Chemistry of Geothermal Waters. – Geyser Area Complexes. – Environmental Aspects of Geysers. – Temporal Changes in Geyser Activity and Their Causes. – Man's Influence on Geyser Activity. – Practical Uses of Geothermal Fluids. – Appendix: Geologic Time (Stratigraphic Column). – Chapter References. – Bibliography. – Index.

This book presents a comprehensive and systematic account of the geological aspects of geysers and related geothermal phenomena. Richly illustrated, it emphasizes their hydrologic and geologic settings and structures, while describing their mode of operation and interaction with the environment. Discussions center on the actions occurring within idealized columnar and pool geysers; current geyser theories are evaluated. The extent and magnitude of geothermal resources are also discussed, including the gases and minerals found in geysers and the chemistry of the water they contain. **Geysers and Geothermal Energy** is unique not only in the depth of its coverage of these fascinating geological and geophysical phenomena, but also in its account of their world-wide occurrence. Geysers and geothermal structures are no longer just of local geologic interest; an understanding of them is important for our increasing use and search for geothermal energy. This volume will provide both geologists and interested students with an excellent introduction to the state of the art in geyser research.

Springer-Verlag
Berlin
Heidelberg
New York

Terrestrial Heat Flow in Europe

Editors: V. Čermák, L. Rybach

1979. 151 figures, 47 tables, 1 twelve-color map.
VIII, 328 pages
ISBN 3-540-09440-7

An understanding of the deep temperature distribution in the earth's crust as well as of the outflow of heat from its interior is of fundamental importance to all earth sciences. This monograph summarizes the most recent results of heat flow studies made in Western and Eastern Europe and discusses the manifold interrelations between geothermics and other geophysical and geological phenomena.
The book's twelve-color map shows the surface heat flow pattern of Europe as a whole in the form of isolines. Geothermal areas are characterized by elevated heat flow; thus, the map will help in identifying suitable geothermal resources and aid in their development. It represents the first undertaking of this magnitude ever attempted.
With more than 3,000 entries on heatflow, this monograph will serve as a manual for the interpretation of the complex geophysical, geological and tectono-physical aspects of crustal and lithospheric structures, of deep-seated phenomena and of tectonic evolution on the European continent.

G. Buntebarth

Geothermie

Eine Einführung in die allgemeine und angewandte Wärmelehre des Erdkörpers

Hochschultext
1980. 64 Abbildungen, 11 Tabellen. IX, 156 Seiten
ISBN 3-540-10423-2

Inhaltsübersicht: Einleitung. – Physikalische Grundlagen zur Wärmeleitung. – Thermische Eigenschaften von gebirgsbildenden Gesteinen. – Analytische Behandlung von konduktiven Temperaturausgleichsvorgängen in der Erdkruste. – Der thermische Zustand des Erdinnern. – Methoden der Temperaturermittlung. – Erdwärme als Energiequelle. – Anhang. – Literaturverzeichnis. – Sachregister.

Springer-Verlag
Berlin
Heidelberg
New York